2003

Damaged Ecosystems and Restoration

Damaged Ecosystems and Restoration

Editor

B C Rana

Sardar Patel University
India

World Scientific
Singapore • New Jersey • London • Hong Kong

Published by

World Scientific Publishing Co. Pte. Ltd.

5 Toh Tuck Link, Singapore 596224

USA office: Suite 202, 1060 Main Street, River Edge, NJ 07661

UK office: 57 Shelton Street, Covent Garden, London WC2H 9HE

British Library Cataloguing-in-Publication Data
A catalogue record for this book is available from the British Library.

First published 1998
Reprinted 2003

DAMAGED ECOSYSTEMS AND RESTORATION

ISBN 981-02-3174-1

Printed in Singapore.

PREFACE

The planet Earth which houses us along with innumerable other features is supposed to be the best creation of Nature. James Irwin, a cosmonauts, after his voyage in space said, " The Earth reminds us of a Christmas tree ornament hanging in the blackness of space..." Mother earth has given us a lot without expecting any thing in return. Fortunately for us and unfortunately for Nature, we are different from other animals in having emotions and a lot of gray matter power. We, the human beings always have a craving for controlling and dominating the phenomena confronted by us. In the process we have taken Nature for granted for too long. In our ignorance we have forgotten that natural resources are not limitless. The situation is worsening: "Mother earth will never seem the same again...", "Man is making deserts of the earth...", Earth's wrath at our assaults is slow to come, but relentless when it does...." Very few will deny that 'Anything in excess is dangerous ? it is needless to say that our craze for possession has caused total environmental mess. I will not like to go in details of the things and pen them down here, as these have been the worry of not only the common people but also all the levels of media. There is no point in making hue and cry at this juncture about the environmental dismantling. We have to remember that now we are to deal with cards (play cards), we are no body to frame them, yet we have to play skillfully with whatever poorest cards in hand and win the game. This collection of articles show some of the clues of playing such cards and march towards remedy if not victory.

Restoration of damaged ecosystems is given priority in environmental science and technology. But we should keep in mind that preventing damage is less expensive than restoring damages. The vital question is: What are the prospects for repairing damages we have done to various natural resources? Articles here, I believe will provide effective answer to this query.

An overview : Several environmental issues are of prime concern for human welfare. Sustainable ecosystem is essentially an important issue. Ecosystems at all levels have been damaged severely because of intentional exploitation and careless management of natural resources. This has jeopardised the ecosystem services (ecosystem services are those functions of natural systems perceived as beneficial to human society) so essential for sustainability of life on this planet. Once ecosystems are damaged they provide diminishing services. To avail these services continuously in sustainable form it is essential that ecological capital should not be destroyed. Restoration requires recreating of both structural and functional attributes of damaged ecosystems. However restoration can not be perfect and predisturbed conditions may not be achieved. If we are to achieve sustainable use of the planet, at some point, the rate of ecological repair must equal the rate of ecological destruction, so that no net loss situation can be achieved. Ecological restoration is a positive statement of co-operation with natural systems. In eco-societal restoration and sustainability of human society's life-support system Cairns has highlighted eight classes of restoration and various obstacles to the restoration process. It is the process of re-examining human society's relationship with natural system so that repair and destructions are balanced. Further, Heckman and Cairns have emphasised that process of true ecological restoration should not be only problem targeted but it should be in tune with the complex dynamics of ecosystem functions. Reversing the process of deterioration, although provide many positive benefits, negative features like high cost and uncertainties regarding technologies to be employed should also be considered.

Gray, has described an interesting clean up story of damaged sites of Handford and Pantex Plant of USA. Both these sites for long were used for nuclear and non-nuclear weapons assembly and disassembly and were contaminated by radioactive and hazardous explosive chemicals. The radiological, chemical and biological monitoring of damaged sites for many years have revealed useful information as an ecological experimental site for the restoration process of damaged ecosystems.

Deforestation is a major ecological problem in many parts of the world. It severely affects soil erosion, soil fertility, biodiversity carbon and hydrological cycles. Tropical moist forest of Latin America has been the major target and very large area has been cleared for agricultural and other purposes. Holl, in her attempt to restore such a complex ecosystem has described a number of factors which limit forest recovery and has also highlighted the strategies for quick recovery. The positive efforts for restoration may help stop more destruction and provide future hope for better management of such forests.

The disposal of hazardous wastes has become an important environmental problem. There have been a widespread increase in domestic and industrial pollution. As a result of advances in industrial technology and intensification of agricultural practices various new xenobiotic, organic and inorganic compounds have been produced. Halogenated compounds like DDT, BHC, Dieldrin, Chlorodane are widely used in agriculture and industry. The recalcitrance, toxicity and some time carcinogenecity of these compounds have led to the wide spread concern. Similarly heavy metals are non-biodegradable and once released into the environment become integral part of the habitat and persists for very long time. Disposal of domestic wastes and effluents contaminated with inorganic nutrients like nitrogen and phosphorus are the major causes of eutrophication in water bodies. Manufacturing of dyes and dye intermediates and discharge of huge dye wastes in water bodies have polluted many water bodies in many developing countries including India (number of chapters have dealt these aspects).

A good progress has been made in the treatment of various types of wastewaters. Large number of physical, chemical, biological and biotechnological methods have been developed for restoration of various damaged ecosystems. Biological methods for wastewater treatment has considerable potentials. The use of algae, cyanobacteria, bacteria, fungi, mosses, macrophytes and other aquatic plants play a critical role in the decomposition of sewage and wastewater, and consequently restore the aquatic environment. Biotransformation and environmental remediation techniques are used in the modern biotechnology. Immobilised cell technology attracts considerable attention because of many advantages. Immobilised algal biotechnology is increasingly used for removal of metals and other hazardous chemicals (Pradhan, Singh and Rai). Over the past decade constructed wetlands are used for the treatment of wastewaters. Free floating hydrophytes treatment system have been used for treatment of municipal wastewater, agricultural and industrial waste waters (Ho and Tsang). Natural wetlands are unique ecosystems. Their ecological and economical role was only recently recognised. Efforts are being made on the global level to protect and conserve wetland ecosystems. Chaudhari and Sarkar, in their paper have described in detail the principles of wetland ecology, factors threatening wetland ecosystems, their conservation and ecomanagement strategies both at global and local levels.

Dyes are biodegradable by only few selected micro-organisms which include selected strains of bacteria, actinomycetes and fungi. A promising new area in environmental biotechnology is the use of radical generating enzyme system to degrade pollutants, have been discussed by Patel and Rawal. Many technological options are available to meet the challenges for maintaining the ecological balance. Among them biotechnological processes play very important role in prevention and restoration of damaged ecosystems. The available microbiol technological options for maintaining ecological balance, wastewater biotreatment, bioremediation and biofilteration technologies are highly promising. These strategies have been dealt well in details by Desai and Dave, and Madamwar and Gandhi. With the recent breakthrough in genetic engineering it is now possible to produce tailor made micro-organisms with desired degradative capabilities. Recombinant DNA technology by-passes all the constraints on genetic exchange between two distinct species. Many novel approaches have been developed for the treatment of toxic or recalcitrant wastes by developing specialised bacteria or enzymes for such wastes.

India's heritage of culture, religion, philosophy and ethics are age old. Reddy has traced the Indian ecological ethics, customs, morals and traditions which were in harmony with nature and its principle and ask why can't we respect the merit of these ecological ethics for restoring our ecosystems.

The contributors of the articles, I must say have wide experience in the field of environmental problems and have some first hand clues to offer to mankind for their enhancement of judicious look at Nature. I am sure the book will interest all those who have respect and concern for environment.

I wish to thank all the contributors for their comprehensive chapters. I am grateful to Prof. John Cairns Jr. whose inspiration and encouragement made me to take up this task. My special thanks are due to Profs. J. D. Patel and D. S. Mishra for their help. I am indebted to my well wishers, authorities in Sardar Patel University and my family members who helped and encouraged me in completion of this work. I am thankful to the publisher who has readily accepted to take up this work for publication. My thanks are also due to Mr. Saket Amin and Amit Upadhyaya who made the task of preparing the manuscript easy for me.

B C Rana
12 April 1997
Vallabh Vidyanagar, India

List of Contributors

John Cairns Jr.	University Distinguished Professor of Environmental Biology and Director University Centre for Environmental and Hazardous Materials Studies Department of Biology, Virginia Polytechnic Institute and State University Blacksburg, Virginia 24061-0415 USA
Karen D. Holl	Environmental Studies Board University of California, Santa Cruz, CA 95064 USA
Robert H. Gray	Manager, Environmental Protection Battelle, Battelle Pantex, P.O. Box 30020, Amarillo, Texas 79177 USA
John R. Heckman	Department of Biology, Virginia Polytechnic Institute and State University, Blacksburg, Virginia 24061-0415 USA
Y. B. Ho	14718-57th Avenue South TUKWILA, Nr, Seattle WA 98168 USA
J. S. H. Tsang	Department of Botany, The University of Hong Kong, Pokfulam Road, Hong Kong
P. R. Chaudhari	Environmental Biotechnology Division, National Environmental Engineering Research Institute NEERI, Nagpur 440020 India
Rekha Sarkar	Environmental Biotechnology Division, National Environmental Engineering Research Institute NEERI, Nagpur 440020 India
Subhashree Pradhan	Laboratory of Algal Biology, Department of Botany, Banaras Hindu University, Varanasi 221005 India
Sarita Singh	Laboratory of Algal Biology, Department of Botany, Banaras Hindu University, Varanasi 221005 India
L. C. Rai*	Laboratory of Algal Biology, Department of Botany, Banaras Hindu University, Varanasi 221005 India (For Correspondence)
M. V. Martinova	Moscow Water Problems Institute, Novo Basmannaja 10 Moscow Russia 107078
John H. Hartig	Environmental Scientist, International Joint Commission, 100 avenue Ouellette Avenue, Windsor, Ontario N 9A 6T3 Canada

Datta Madamwar	Microbiology Laboratory, Department of Biosciences, Sardar Patel University, Vallabh Vidyanagar 388120 India
Manish Gandhi	Microbiology Laboratory, Department of Biosciences, Sardar Patel University, Vallabh Vidyanagar 388120 India
B. C. Rana	Department of Biosciences, Sardar Patel University, Vallabh Vidyanagar 388120 India
Jitendra D. Desai	Environmental Science and Applied Biology Division, Research Centre, Indian Petrochemicals Corporation Limited, P. O. Petrochemicals, Vadodara 391346 India
Heena Dave	Environmental Science and Applied Biology Division, Research Centre, Indian Petrochemicals Corporation Limited, P. O. Petrochemicals, Vadodara 391346 India
K. C. Patel	Microbiology Laboratory, Department of Biosciences, Sardar Patel University, Vallabh Vidyanagar 388120 India
Bhavin Rawal	Microbiology Laboratory, Department of Biosciences, Sardar Patel University, Vallabh Vidyanagar 388120 India
V. Sankaran	Department of Botany, NGM Autonomous College, Pollachi 642001 India
C. M. Joy	Department of Botany, Sacred Heart College, Thevara, Cochin 682013 India
Ammini Joseph	School of Environmental Studies, Cochin University of Science and Technology, Cochin 682016 India
A. S. Reddy	Department of Biosciences, Sardar Patel University, Vallabh Vidyanagar 388120 India

Contents

ECO-SOCIETAL RESTORATION : REHABILITATING HUMAN SOCIETY'S LIFE SUPPORT SYSTEM

JOHN CAIRNS, JR.

Department of Biology
Virginia Polytechnic Institute and State University
Blacksburg, Virginia 24061-0415 USA

1 Introduction

Since the agricultural and industrial revolution began, the rate of ecological destruction has far exceeded the rate of ecological repair. If we are to achieve sustainable use of the planet at some point, the rate of ecological repair must equal the rate of ecological destruction so that a no-net-loss situation can be achieved. The National Research Council[1] recommends that ecological restoration be carried out at a landscape level wherever possible. This recommendation is by no means a denigration of small restoration projects, but rather is intended to point out that larger restoration projects have a higher probability of becoming self-sustaining and economies of scale can be considered. Furthermore, most landscape level systems are a mosaic or patchwork of smaller ecotypes. Each of these is likely to function best under somewhat different conditions and to endure varying types of stresses differently. This diversity means that the vulnerability of a particular ecotype does not mean that the entire restoration effort is lost but, rather, just a component thereof. Finally, when restoration projects are undertaken, the expectation is that they will endure for a considerable time thereafter, or the investment made in the restoration effort would not be fully recovered or, more appropriately, would not "pay dividends." But increasing the spatial and temporal scales of any project vastly increases the likelihood of substantive interactions with human society. To put it more bluntly, no landscape level restoration effort will endure for long periods of time without an endorsement of its dynamics and benefits by the human society with which it is associated. This understanding will require a higher level of ecological literacy on the part of legislators, the business community, regulators, ant the general public than now exists.

The multi-dimensional undertaking of eco-societal restoration will necessarily involve considerable time periods. Also, while restoration projects almost always are ecological far superior to the damaged condition, the precise outcome is usually uncertain. However, the reasons for integrating restoration into societal activities far outweigh the expense and uncertainties[2]. Practicing ecological restoration significantly contributes to the expansion of scientific knowledge and ability, although very few people presently practice ecological restoration. While, in theory, any ecologist should be suited for restoring damaged ecosystems, most carry out their research on populations of single species with infrequent attention to chemical/physical parameters[3]. Additionally, ecological restoration involving society requires a far broader approach than the field of ecology as it is presently practiced.

Having a number a number of ecosystems that are being repaired will provide more natural sources of recolonizing species to other damaged ecosystems. These systems can also provide additional sources of recolonizing species where transport is necessary. Since restoration to predisturbance condition is not often a viable option[4], creating naturalistic assemblages of plants and animals, consisting of species not necessarily there before the disturbance, will provide both scientific evidence and a number of case histories that both the general public and professionals can examine when making choices of restoration end points.

Current ecological restoration activity also contributes to an eventual decrease in the cost of restoration. The ability to estimate the cost of restoration will be markedly improved as the number of projects increases. When the full cost of ecological restoration is better documented, it may well act as a deterrent to further damages because the dollar costs can be incorporated into comparisons of alternatives actions. These full cost numbers will also enable the amount of money in restoration bonds to be determined with more precision.

Ecological restoration projects in each ecoregion, and preferably in each major area of the country, will provide demonstrations for local citizens, and this will vastly increase environmental literacy. On a more general level, however, human society's practices are the best indication of its ethos or set of guiding beliefs. Ecological restoration is a positive statement of cooperation with natural systems. Preserving those systems still undamaged and protecting those restored would be an even more positive statement, especially if accompanied by major restorative efforts for presently damaged systems.

2 Classes of Restoration

Cairns[5] describes eight classes of restoration, representing the different levels of restoration that may result from a matrix of both ecological and social criteria. The multivariate natural of ecological damage creates a wide variety of situations that must be addressed. This classification system provides a framework in which damaged ecosystems can be compared and decisions concerning restoration can be made.

2.1 Class A - Restoration to an Ideal State

Ecological restoration to an ideal state involves at least two choices : (1) resetting the ecological clock to the predisturbance condition or (2) resetting the ecological clock to the ecological condition that would be characteristic of the system had no ecological disturbance occurred. Both of these choices are difficult because information about the condition of the ecosystem before it was disturbed may not be available in sufficient detail or depth to replicate or repair the system. Even more difficult is determining where the system would be had no disturbance occurred. As a consequence, even if it is possible to agree upon goals with both ecological and human values, the information base may not be adequate to do so. Ideally, Class A restoration should be self-maintaining in the sense that no ongoing management cost will be involved.

2.2 Class B - Restoration of Selected Attributes

If the environmental literacy of the society involved with the damaged ecosystem is sufficiently high, it may be possible to repair damaged components of recreational, commercial (e.g. fisheries), societal (e.g. wetlands useful in storing flood waters), or aesthetic value. It is not clear whether Class B restoration will be self-maintaining. Almost certainly, whether or not the system will be self-maintaining will have to be decided on a case-by-case basis and, until ecological restoration becomes more common, a fair amount of uncertainty will be involved. This situation means management costs for ecological monitoring will be incurred to determine if, in fact, self maintenance is a reality. In Class B restoration and all levels of restoration that follow, assignment of management and maintenance responsibility is extremely important. Especially important is the determination of an ongoing financial base. When choosing different levels of restoration, it is extremely important that both the cost for restoring and cost for maintaining over substantial temporal periods are at least estimated and part of the financial planning. Class B restoration might be utilized where an oil spill or some other kind of ecological damage has occurred and the probability of recurrences of problems of one sort or another is fairly high.

2.3 Class C - Restoration to an Alternative Ecosystem

In some cases, as in surface mining, the ecosystem may be so disrupted due to alteration of the soil profiles, slope, removal of vegetation and topsoil, and the like that restoration to predisturbance condition or even an approximation thereof is either too expensive, highly improbable for scientific reasons, or not high on the social agenda. In this case, a naturalistic assemblage of organisms having desirable attributes, but not a close match to the predisturbance community, might be installed. In some cases, the constructed ecosystem (or created ecosystem) might be dramatically different from the predisturbance ecosystem. For example, 117,000 acres of wetlands have been lost in the United States over the last approximately 200 years[1]. In some states, such as California, the losses are as high as 91%, in others, only a few percent. Replacement of these lost wetlands, although not in their original geographic location, might be quite desirable to save endangered species, to furnish habitat for migratory birds and other organisms, to serve some of the functions that wetlands can perform well in the new situation, and a variety of other reasons.

2.4 Class D - Natural Recovery

Years ago, when ecological restoration was beginning to be a major part of my professional life, a fly ash pond from a coaldriven steam generation plant damaged a substantial section of the Clinch River in southwestern Virginia, U.S.A. My graduate students at that time and I focused entirely on natural processes and made no attempt whatsoever to assist the recovery process. This spill occurred in an area with numerous tributaries that were unaffected; the head waters of the Clinch River were also unaffected. Industry or other development in the area was spares and, except for a few groups such as mollusks, the system recovered with amazing rapidity. Possibly this was not only because invading organisms quickly recolonized, but because there was no residual toxicity. Even when natural recovery occurs with colonists from undisturbed adjacent natural ecosystems, the final result may not be an exact replication of the predisturbed ecological condition. Furthermore, in areas where sources of invading species are small and natural transport mechanisms may not function well because of ecological barriers, natural recovery is probably not going to occur in time spans acceptable to human society. Nevertheless, it should not be entirely ignored.

2.5 Class E - Restoration of Hazardous Waste Sites

The literature on the restoration of hazardous waste sites is extremely limited, and much of the information is not in the professional literature but rather in high security or limited distribution documents. Nevertheless, persistent highly toxic wastes are extremely difficult to eliminate in situ and, therefore, restoration to any ecological level that will permit use by human society without risk is not likely for a number of years. Nevertheless, if a hazardous site is decommissioned and no longer involved with the generation of hazardous waste but merely the storage thereof, a number of options are available even if the site is going to have restricted use. Ecological improvements resulting in the following should prove both economically and ecologically attractive: (1) minimize run off, (2) minimize contamination of groundwater, (3) enhance biotransformation of degradable substances, (4) exert low invasion and colonization pressure on adjacent ecosystems, (5) be outcompeted easily by more suitable organisms once the presence of hazardous materials has been reduced significantly, and (6) serve as early warning systems or biological sentinels to prevent export of deleterious materials to neighbouring ecosystems. Obviously, maintenance and management costs, as well as environmental monitoring costs, of such a limited restoration project would be much higher than the other options already listed.

2.6 Class F - Restoration in an Area with a High Risk of Ecoaccidents

Even though it seems quite rational not to restore to the predisturbance condition in an area with a probability of ecoaccidents, a counterview is that this is an invitation to pollute. Until human society recognizes it is dependent on both a technological and ecological life support, this argument has merit. If, however, society uses opportunity cost analysis[1] in deciding what damaged ecosystems merit restoration and where the money can be spent most wisely, attempts to restore to the predisturbance condition in high risk areas are likely to occur. If resilient species (i.e. either resistant to or recover quickly from anthropognenic stress) are selected for the restoration, the ecosystem services delivered by the ecosystem will be less certain.

2.7 Class G - Restoration after Contamination by Genetically - engineered Organisms (GEOS)

Some readers may justifiably feel that neither Class G nor Class H restoration qualify as true ecological restoration. However, since genetically- engineered organisms have the potential to displace indigenous organisms, they might well-affect both the structure and function of an ecosystem. The cause of the damage may be new, but all ecological damage should be repaired regardless of the cause. One of the best examples of exotic organisms causing ecological damage is given in[6] and a more recent publication on this subject is Cairns and Bidwell.[7] Class G Restoration might include elimination of the GEOS, followed by reconstruction of the damaged ecosystem to either the predisturbance condition or one of the alternative classes of restoration just discussed.

2.8 Class H - Restoration Designed Primarily to Protect Adjacent Ecosystems

This type of restoration is especially suitable for hazardous waste sites with long-term persistent risks. The restoration would be designed to immobilize wastes and prevent their biological transport and concomitantly aid in their transformation to less hazardous materials. If one looks on ecological restoration from a landscape perspective, this alteration of an ecological hazardous waste site to reduce the risk to the ecological landscape as a whole (e.g. adjacent and distant ecosystems) might well qualify as restoration. Whatever terminology is used, this is an increasingly essential activity for human society and will continue to be so as long as wastes are produced that remain hazardous for long periods of time and that cannot be readily assimilated into natural ecosystems.

3 Obstacles to Ecological Restoration

(1) Science is not sufficiently robust to predict accurately the consequences, if any, of alteration of major environmental alterations nor is it well-prepared for restoring damaged ecosystems !

So little is known about the consequences of the destruction of tropical rain forests, loss of topsoil at a rate greater than replacement, or damage to the ozone layer that, when evidence that will persuade the doubters appears, it may be too late to save the damaged ecosystems. On the other hand , evidence in the literature[1] shows that practically every effort to restore damaged ecosystems results in an improvement, if not a return to, predisturbance conditions. Therefore, learning by doing can continue with the expectation of a high probability of significant ecological improvements for each restoration project. In addition, another valid expectation is that the longer restoration projects are carried out, the more robust the base of scientific information and qualified practitioners will become. Human society may by using species from relatively undamaged ecosystems and inadvertently impairing their integrity, but, again, no course of action is free of risk, including doing nothing. However, the learning process will be enhanced by well-designed restoration efforts.

(2) Human population growth and individual expectations of increased affluence are so well-established and have so much momentum that nothing can reverse these trends !

The fatalistic assumption that neither individuals nor human society can change sufficient enough to prevent destruction of the ecological portion of Earth's life support system may well be correct; however, there is sufficient evidence of the ability of human society to change that it seems irresponsible not to do everything possible to repair the ecological portion of human society's life support system. Even in a worse possible case situation, more people will enjoy a higher quality life for a longer period of time than is likely if nothing is done.

(3) Humans cannot destroy every species on the planet, and natural systems have recovered from past great extinctions.

If seems virtually certain that humans cannot destroy all nondomestic species without having a dramatic, possibly fatal, effect upon human society itself[8]. A return to the pre-industrial, pre-agricultural level of biodiversity is, in terms of the temporal framework of human society, essentially meaningless.

Since the publication of the Bruntdland Report[9], discussion has focused on such terms as *sustainable development sustainable growth* and *sustainable use of the planet,* which seems to be a way of ensuring that future generations are not deprived of resources or a functioning ecological life support system. The use of the terms *growth* and *development* imply that present human societal behaviour and practices can be continued indefinitely by associating them with the word *sustainable.* But, this is not a rational position to take in the long term. Cairns[8] speculates that a coevolutionary interaction exists between human society and natural systems in that each can affect the other beneficially or adversely. Abundant evidence shows human society's adverse effects upon natural systems. It is not clear to most members of human society that natural systems can have adverse effects upon them as well. Garrett[10] provides a persuasive analysis that environmental disequilibrium is facilitating the spread of plagues. Cairns[11] and Cairns and Pratt[12] discuss ecosystem services and define them as those functions of natural systems perceived as beneficial to human society. Arguably, if human society were more environmentally literate, practically all ecosystem functions would be viewed as services (Table 1).

Table 1 : Illustrative ecosystem services and their underlying ecological functions.

Ecosystem service	Related ecosystem function	References
Biomass production for food, building materials, and fuel	Capture of solar energy through photosynthesis, regeneration of essential nutrients and soils	Cairns[20], Ehrlich and Mooney[21], Hitzhusen[22], Izac et al.[23], Vitousek[24]
Assimilation of pollutants	Decomposition, detritus processing	Cairns a, b[11,12]

Flood control, water purification, water transportation	Maintenance of the hydrological cycle	Van Wilgen et al[26], Vitousek[24], Culotta[27]
"Pest" species control	Predator/ prey relationships with insectivorous birds, insects bats, etc.	Cairns b[25], Ehrlich and Mooney[21]
Provision of new compounds for medicinal and other uses	High biodiversity through competition, selection, and speciation	Cairns b[25], Janzen[28], Vitousek[24], Davis et al.[29], Lugo[30]
Maintenance of a breathable and protective atmosphere	Biogeochemical interactions	Cairns[20], Ehrlich and Mooney[21], Vitousek[24]
Aesthetic and spiritual values	All natural functions	Cairns[20], Callicott[31], Power[32]

(4) US versus THEM

Quinn[13] and others have noted that many members of human society believe that humans are apart from nature rather than a part of nature. Yet, if one structures the issue in terms of sustainable use of the planet (discussed at length later), all parties share a common destiny and the polarization does not contribute to the solution of the problem.

Quinn also contrasts two groups in human society -- the takers and the leavers. Takers are people who exploit natural resources in such a way that the prospects for sustainable use of the planet for future generations is greatly diminished. Leavers attempts, and sometimes succeed, in preserving ecological capital (e.g., topsoil) so that the replenishment rate is at least equal to the rate of loss; in other words, a balance between loss and replenishment. To the two groups mentioned in Quinn's book *Ishmael*[14], I would like to add a group called healers. The healers (those engaged in ecological restoration) add to existing ecological capital by repairing some or most of the damage done by the takers. Sometimes the takers are forced to pay for this ecological repair but not for the loss of ecological services that occurs because the ecosystems have been damaged. Nevertheless, the discussion of sustainable use of the planet now underway requires that society diminish the damage caused by the takers, increases the legacy of the leavers, and ideally move toward healing the damage human society has caused so that future generations benefit from present activities rather than be impoverished by them. In *The Ishmael Companion*[15], Williams[16], notes that "takers believe in their revolution, even when they enjoy none of its benefits. There are no grumblers, no *dissidents,* no counter revolutionaries." In short, the taker philosophy is still dominant, even though many of the takers do so for very understandable reasons, such as poaching endangered species to provide themselves and their families with a standard of living somewhat above subsistence or at least above their nonpoaching contemporaries.

4 Developing Principles for Priority Setting and Decision Making

It is *sine qua non* that all priority setting and decision making must be based on well-established principles if both priorities and decisions are to be communicated, understood, and, most importantly, supported by human society. In the United States and most other countries, responsibility for components of the environment, both technological and ecological, have been fragmented and isolated from each other. Establishing a high priority for ecological restoration will require an enormous broad-based improvement in ecological and environmental literacy.

4.1 Principle Number 1

Principle and programmes for ecological restoration should emphasize landscape perspectives. Fragmenting the landscape rather than emphasizing a systems approach is neither cost effective nor likely to result in sustainable use. Fortunately, technology now permits viewing landscapes much more effectively than it did even a few years ago. Probably the view of planet Earth from space has been the most dramatic and effective means of promoting the landscape perspective.

4.2 Principle Number 2

Ecological restoration policies and individual projects should be designed and executed according to the principles of adaptive planning and management. Nevertheless, if society waits until a robust scientific and social base has developed, it may be too late to restore badly damaged parts of the planet.
Each restoration project is unique, and uncertainties are likely to be significant, although not project threatening. Even when a wetland or forest restoration is being carried out in an ecoregion or an ecosystem where seminal projects have already been completed, uncertainties will still exist because of the vagaries of weather, air pollution, and the like. Those engaged in the restoration project, if not research scientists, wish to use as little professional judgment as possible because the more used, the more responsibility of those in charge. Therefore, while flexibility and adaptive planning and management initially seem enormously attractive, in practice, they usually are not perceived as such. Even the general public would like to have assurances regarding the outcome. This mindset is particularly true in the present climate where research in universities and colleges is viewed by legislators and those who provide funding as a threat to "good teaching." Problem solving that involves faculty and students analyzing a problem and discussing alternatives in the face of uncertainties is the finest form of teaching because it involves analytical and thought processes and reasoning rather than the type of learning where "facts" are passed on to students in a formal lecture setting and later regurgitated during exams.

Simply observing a restoration project is not enough, monitoring is required. Measurements must be relative to explicitly stated goals, and the information must be translated into restoration policy and programme redesign throughout the project's existence.

4.3 Principle Number 3

Evaluation and ranking of restoration alternatives should be based on an assessment of opportunity costs rather than on the traditional benefit-cost analysis. The formidable challenge in ecological restoration management is to evaluate trade-offs, not only between alternative approaches to restoration. Because different structures and functions of any ecosystem yield different values or services, choosing whether and how to restore usually amounts to choosing one set of values or services over another. The appropriate basis for defining ecosystem values or services is a central, analytical question. Traditional cost-benefit analysis is more programmed to one alternative on one site than an array of options and alternatives. A more extended discussion of opportunity cost-analysis may found in National Research Council[1] .

5 Restoration Policy

No ecological restoration project at the landscape level can be carried out in isolation from human society. In addition, human society will not likely alter behavior and practices when necessary without sufficient knowledge to appreciate fully the benefits of the restoration undertaking. Most likely, ecological restoration activities will have to be integrated with other activities of human society[1] . If one accepts these assumptions, it is abundantly clear that no ecological restoration activity will be successful or endure for substantial periods of time unless policy positions are developed and affirmed by human society. The first step in the policy process is the establishment of national restoration goals. (The following section has been condensed from National Research Council volume[1] with permission.)

5.1 Establishing National Restoration Goals

Although the National Research Council[1], book focuses on aquatic ecosystems and emphasizes a landscape approach, which necessarily includes the terrestrial system associated with the aquatic ecosystems with which they are intimately involved in the hydrologic cycle, the goal statements given here have been modified to include all ecosystems, including oceans. It is essential to coordinate and integrate national policies through the United Nations Environmental Programme or some equivalent organization. Within each nation, subdivisions such as provinces or states should also be involved, but their policies must necessarily be congruent with national policies and national policies should be congruent with global objectives. For example, migratory birds, air, and hydrologic cycle do not pay close attention to political boundaries. Therefore, the restoration goals in one nation will not be entirely effective unless the global ecological landscape is considered.

5.1.1 Goal Number 1

A national restoration strategy should be directed toward broadbased and measurable goals.

Much attention[9,17,18] but little action has surfaced toward a sustainable future. Sustainability essentially means acting in such a way that future generations will have a quality of life equal to or better than the present. Not depriving future generations by present actions is the essence of sustainability. Stated more bluntly, the present generation should not be stealing ecological capital from its children, grandchildren, great grandchildren, and generations to follow.

Restoration goals could be to have a no-net loss of ecosystem types such as wetlands, to restore ecosystems at a rate greater than they are being destroyed, or to continue present practices of restoring them less rapidly than they are being destroyed.

Goals should have both spatial and temporal dimensions! An illustrative example from the National Research Council[1] concerns restoring wetlands in the United States at a rate that offsets any further loss of wetlands and contributes to an overall gain of 10 million wetlands acres by the year 2010, largely through reconversion of crop and pasture land and modification of existing water control structures. This number represents <10% of the total number of acres of wetlands lost in the last 2000 years (estimated at 117 million acres). The NRC[1] book recommends that a total of 400,000 miles of streams and rivers be restored within the next 20 years.

Unquestionably, even establishing any restoration goals, including a rate of repair lower than the rate of destruction, will produce a contentious debate in human society as a whole and in those countries where debate is possible. It is, at present, extremely difficult to forego present "needs' for the sake of future generations, who are invisible to everyone, but this sacrifice is inevitable unless we destroy the planet as a viable place for human society. Similarly, compassion is not easily felt for species other than our own, particularly those that are unlovable. Even compassion for lovable species, such as the panda and koala, may quickly give way to human progress and growth of society. Compassion is usually the result of a high level of literacy (perceiving and understanding the needs of others), coupled with an ethical and moral sense. Restoration ecology and sustainable use of the planet will be the acid test in these categories for human society and will undoubtedly determine the future quality of life and, possibly, the continued existence of human society itself.

5.1.2 Goal Number 2

A national ecosystem restoration assessment programme should be established to monitor the degree to which established goals are being achieved.

In this case, monitoring means the gathering of evidence to determine the degree to which previous goals have been met. There is little point in establishing goals and objectives if the effectiveness of the implementation toward achieving these goals is not based on persuasive, hard evidence. Remote sensing and a variety of other technological achievements should make this task much easier than it would have even been a decade ago. Additionally, the underlying science of ecological restoration is not yet robust nor is the integration of societal activities so well-advanced as to optimize the protection of both the technological and ecological portions of the life support system.

6 Redesigning Both Policy and Programmes of Government Agencies

Each country and each region should modify the following to fit each unique situation, but these general guidelines should prove useful:

(1) A common definition of restoration should be used by all agencies within the governmental structure so that the public and legislators are not confused by a multiplicity of definitions. Cairns[5] has recommended a series of definitions of restoration, each with a different degree of progress toward the predisturbance condition and each with different levels of management responsibilities once the restoration has been completed. Ideally, an ecological restoration should lead to self-maintenance of the restored ecosystem in the sense that all natural systems have been self-maintaining unless disturbed by substantial climatic or human stress. Restoration to a predisturbance condition with the "original" species may be an aspiration rarely achieved (Figure 1) but is, nevertheless, a way of determining the level of restoration for which society is willing to pay and which is scientifically achievable (Figure 2).

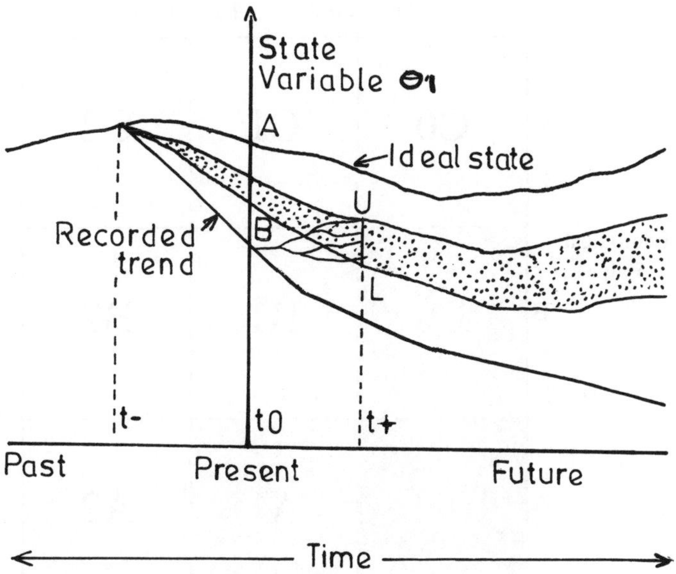

A -	Ideal value of the state vatiable without any human activities t_0.
B -	Present value of the state variable
U -	Best value of the state variable that can be achieved based on present state of knowledge at the completion of restoration project provided no economic constraints.
L -	Least acceptable value of the state variable, at the completion of restoration project
	Achievable Functional Envelope

Figure 1: Schematic representation of a restoration scenario (Examples of state variables include river stage, water temperature, and fish species).

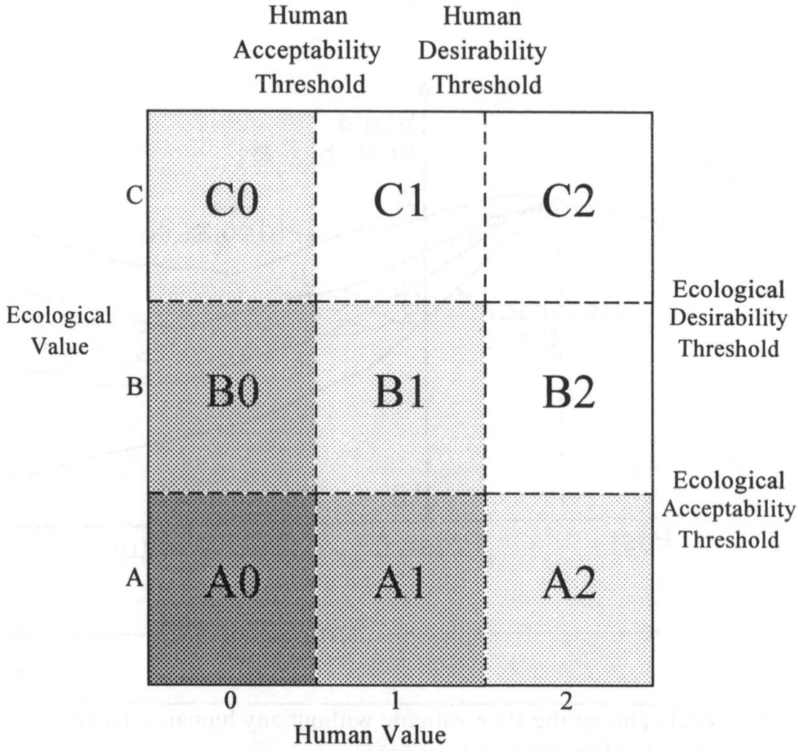

Cell	Ecological Value	Human Value
A0	Unacceptable	Unacceptable
A1	Unacceptable	Acceptable
A2	Unacceptable	Desirable
B0	Acceptable	Unacceptable
B1	Acceptable	Acceptable
B2	Acceptable	Desirable
C0	Desirable	Unacceptable
C1	Desirable	Acceptable
C2	Desirable	Desirable

Figure 2 : Project assessment matrix

Ecosystems are dynamic and constantly changing; therefore, ecosystems are not static although, in some, changes may occur over geological time and not be readily apparent to humans. All of these and many other factors should be kept clearly in mind when developing definitions of restoration so that the term is not misunderstood.

(2) A national restoration strategy should allocate leadership to the central government for landscape level restorations clearly of national significance should rely on local and central units of government to coordinate restoration programmes in regional or local areas. For example, North America's Mississippi River and its tributaries are clearly of national importance, as are the Great Lakes (actually of international importance). The Everglades have been damaged by drainage canals that altered the flow of water. Because the Everglades, although entirely within the state of Florida, is of national significance, the government should bear some significant portion of the financial burden and responsibility for effective restoration activities.

(3) The national government should initiate and implement interagency planning and collective actions so that the fragmented responsibility of the environment for different components of the environment does not impede restoration nor further damage ecosystems. Most government agencies are allotted a specific amount of money for a specific task. Unfortunately, the task of one agency might conflict with another. The U.S. Army Corps of Engineers might be asked to build a dam for flood control which then might interfere with the spawning of migratory fish that reside mostly in salt water but spawn in freshwater or the practices recommended by the U.S. Department of Agriculture might result in increased sediment runoff, or construction might lead to an increased percentage of impervious surfaces, such as roads and parking lots, that might change the runoff patterns so that erosion of stream banks is increased, increasing transport of solids. When solids settle out far down stream, the U.S. Army Corps of Engineers spends money to dredge the channels to keep them open for shipping transport. Working to reduce transport of sediments would result in less money spent dredging, even though other costs might be increased somewhat. Additionally, sediment transport reduce the life expectancy of the reservoirs behind dams because these are sediment traps. By having the various agencies integrate their management practices, environmental improvements would often result, together with a reduced total expenditure of funds. Cairns et al.[19] discusses the implementation of integrated environmental management.

(4) The national government should implement an interagency and intergovernmental process to develop a unified national strategy for ecosystem restoration. This scheme is the only way that integrated environmental management will work effectively and expeditiously. Since management responsibilities for various components of the environment (land, water, and wetlands) are presently fragmented among multiple governmental organizations and between the public and private sectors, merging them effectively will only be possible if there is a unifying theme. As a beginning, establishing common goals and precise legislative language should reduce conflict among the agencies involved. Additionally, requiring common decision criteria and analytical processes will help reduce conflicts originating from this cause. Finally, organization arrangements specifically developed for resolving agency and programme conflicts can be defined in a variety of ways such as legislation, executive orders, administrative rules, or memoranda of agreement or understanding. Naturally, each agency will fight to continue its existence, and integrative environment management may be viewed by many individuals as a threat to job security. Thus, one cannot expect effective integrated environmental management to occur overnight or without funds and directives specifically designed to enhance and facilitate.

(5) Government programmes, both current and proposed, should exploit available opportunities for ecosystem restoration. A number of illustrative examples are to be found in National Research Council volume[1]. In the United States, which has endured severe flooding on the Mississippi and other rivers recently as well as suffering hurricane damage, one might determine whether dwellings and other human artifacats should continue to be built on flood plains where damage is inevitable and whether structures damaged by floods should be repaired or relocated elsewhere. The periodically flooded land could then be restored to its natural place in the hydrologic cycle, almost certainly reestablishing wetlands and a number of other habitat types. In the long run, by reducing risks of damage to structures by flooding, the central government would save considerable amounts of money, not to mention risk to human lives and health. There are many other opportunities at various governmental levels, such as landfills, for ecological restoration.

(6) Each government should establish a national ecosystem restoration trust fund. In the US, many corporations, usually small ones, did much environmental damage and then avoided the financial responsibility by going bankrupt. A way to avoid this, for example in surface mining, is not to issue a permit for mining until a bond has been posted covering the expected cost of

ecological restoration. As is the case for most bonds issued by private insurance companies, the cost is associated with the probability that the purchaser might default, requiring the insurance company to pay the cost of the bond. In short, responsible corporations pay less per $ 100,000 than less responsible ones. In the case of transport of hazardous materials, bonding could be based on the probability of an occurrence of an accident, together with the estimates of ecological risks and hazards involved. As time goes by and appropriate information becomes more readily available, costs can be estimated more precisely; until then, insurance companies will probably err on the high side. Even so, cases will undoubtedly appear when the bond is clearly inadequate, through no fault of the corporation. For example, restoration of a surface coal mining operation might be impeded by a 5-year drought or 5 years of exceptionally heavy rainfall with mud slides, etc., or an exotic species may invade that would upset normal successional processes. In such cases, the governmental restoration trust fund would provides supplemental resources when the bonding resources became exhausted. This could be done with opportunity cost-analysis; the trust fund would allocate available resources where the most ecological "good" would be expected. The trust fund could be established either by direct taxation of each citizen, since ecological restoration presumably is in the interest of all society, or alternatively by placing an additional tax on all organizations and individuals whose activities represent a risk to environmental integrity.

7 Six Ecological Restoration Options Available to Human Society

One of the major difficulties that policymakers have in making a commitment to ecological restoration is the scientific uncertainties involved. For example, will the services provided by natural systems (e.g. maintenance of atmospheric gas balance, water quality, genetic reserves) be adequate even if both population growth and increased levels of affluence are continued that further endanger natural systems ? Because adequate information bases are lacking and there is uncertainty, society is also unaware of the risks involved in not beginning ecological restoration in a major way immediately. Tolerance of scientific uncertainty and tolerance of risks are both proper subjects for debate before decisions are made. However, they are linked -- acting with an intolerance of uncertainty often demands a high tolerance for risk. In short, the consequences of not beginning a major ecological restoration programme for human society may mean that, when there is enough information on the ecosystem services needed, it may be too late because natural systems will not be in good health and in abundance to provide for them. Thus, while uncertainty is reduced, risk is not simultaneously reduced, but, rather, increased if no action is taken.

The six major options open to human society with regard to ecological restoration and the provision of ecosystem services (i.e., the ecological component of human society's life support system) follow:

(1) Continue the present course of action (increased global population growth coupled with increased levels of affluence and consumption per capita) until the consequences of the actions are abundantly clear. This options means that the destruction of ecosystems will continue at a rate much greater than repair with the inevitable reduction of ecosystem services per capita. It may well be that continuing this course of action will condemn ever increasing numbers of people to poverty and that the integrity of ecological systems may be so reduced that they will no longer continue to provide necessary services.

(2) Establish a no-net-loss of ecosystem services, which would mean roughly balancing destruction and repair of natural systems. It should be kept in mind that restoring damaged ecosystems requires much more time than destroying them, so maintaining a balance at any one point in time will mean a considerable exhilaration of ecological restoration. Even if a no-net-loss or equilibrium between destruction and repair is reached and the population continues to grow, a loss of ecosystem services per capita will continue, but the rate of loss would be markedly reduced.

(3) If the global human population continues to grow at its present rate or at significant rate, there would still be a loss of ecosystem services per capita but the rate of loss would be even less than for option 2.

(4) Stabilize human population size globally in concert with a no-net-loss of ecosystem services (i.e. balancing ecological destruction and repair), which would result in maintaining the status in terms of ecosystem services per capita. The first three options worsen the present situation. Option 4 maintains the present situation so that things will be no better nor no worse.

(5) Stabilize human population and exceed no-net-loss. This would mean increased ecosystems services per capita and, thus, provide a greater margin of safety in the delivery of ecological services to human society. In short, there would be a buffering capacity against natural catastrophes, large-scale accidental spills of hazardous materials, or protection against faulty estimations of risks in the use of chemicals or new technologies.

(6) Reduce human population size and increase the rate of ecological repair beyond no-net-loss. This would mean significantly increasing ecosystem services per capita.

Sustainable use of the planet requires exercising at the very least, option 4, and most probably, option 5 and 6. The reason for excluding option 4 as a possible strategy for long-term sustainable use of the planet is that human society is now living on ecological capital, such as old growth forests, and therefore is not really in a sustainable use mode. Option 5 and 6 would be more likely to place human society in a sustainable use mode.

8 Conclusions

Ultimately, human society's relationship with natural systems should be based on an ethos, or set of guiding beliefs, which then must be implemented by both changes in individual and societal behavior and by the skilled gathering and use of scientific and engineering information, as well as information from other disciplines such as economics and sociology. Eco-societal restoration is the process of re-examining human society's relationship with natural systems so that repair and destruction are at least balanced and perhaps, ultimately, restorative practices will exceed destructive practices. Ecological restoration is a positive statement of cooperation with natural systems. Preserving those systems still undamaged and protecting those restored would be an even more positive statement, especially if accompanied by major restorative efforts for presently damaged systems. Ecological restoration is a tangible way of showing respect for the other species with which humans share the planet-restoring the habitats upon which they depend just as much as humans do on their habitat. Of course, many damaged ecosystems will partially recover, perhaps even fully recover on their own given sufficient time. The process can be accelerated with assistance from human society. Ecological restoration is arguably the most significant expression of human society's regard for natural systems because it repairs damage caused by human society, and, hopefully, the restoration process will heighten environmental literacy and concern, thus preventing further ecological damage.

Acknowledgements

I am deeply indebted to John R. Heckman for extremely useful comments on the first draft of this manuscript and its structure. I thank Lisa Maddox for transcribing the first set of alterations for the second draft and Darla Donald, my editorial assistant, for preparing the manuscript for publication.

References

1. National Research Council. *Restoration of Aquatic Ecosystems : Science, Technology, and Public Policy.* National Academy Press, Washington, DC. (1992).
2. J. Cairns, Jr. *American Women in Science Magazine* **23**, 14 (1994).
3. J. Harte, M. Torn and D. Jensen., in *Global Warming and Biological Diversity*. eds. R. L. Peters and T. E. Lovejoy, Yale University Press, New Haven, (1992).
4. J. Cairns, Jr. *Environmental Professional* **11**, 152 (1989).
5. J. Cairns, Jr. *Environmental Professional* **13**, 186 (1991).
6. H. A. Mooney, and J. A. Drake, *Ecology of Biological Invasions of North America and Hawaii.* Springer- Verlag, NY. (1986)
7. J. Cairns, Jr. and J. R. Bidwell *Biodiversity Conservation.* (In Press).
8. J. Cairns, Jr. *Water Science and Technology Board,* National Research Council, Washington, D. C. (1994 b).
9. The World Commission on Environment and Development. *Our Common Future.* Oxford University Press, Oxford. (1987).
10. L. Garrett, *The Coming Plague.* Farrar, Straus and Giroux, New York. (1994).
11. J. Cairns, Jr.,*Human Ecological Risk Assessment* **1**, 171 (1995 a).
12. J. Cairns, Jr. and J. R. Pratt in *Evaluating and Monitoring the Health of Large- Scale Ecosystems,* eds. D. J. Rapport, C. L. Gaudet and P. Calow, NATO ASI Series, Vol. 128. Springer-Verlag, Germany; (1995).
13. D. Quinn, in *Environmental Literacy and Beyond.* eds. B. Wallace, J. Cairns Jr. and P. A. Distler, Virginia Polytechnic Institute and State University, Blacksburg, VA. (1993).
14. D. Quinn, *Ishmael.* Bantam - Turner Books, NY. (1991).
15. S. H. Maher and R. Quinn, *The Ishmael Companion : Classroom Notes from Teachers.* The Hard Rain Press, Austin, TX. (1995).

16. R. Williams, in *The Ishmael Companion : Classroom Notes from Teachers.* eds. S. H. Maher and R. Quinn, The Hard Rain Press, Austin, TX. (1995).
17. R. F. Weston, *The Weston Way* **21** : 29 (1995).
18. S. R. Ingman, X. Pei, C. D. Ekstrom, H. J. Friedsam, and K. R. Bartlett, *An Aging Population, an Aging Planet, and a Sustainable Future.* Texas Institute for Research and Education on Aging, University of North Texas, Denton, TX. (1995).
19. J. Cairns, Jr., T. V. Crawford, and H. Salwaseer, *Implementing Integrated Environmental Management.* Virginia Polytechnic Institute and State University, Blacksburg, VA. (1994).
20. J. Cairns, Jr. *Association Southeastern Biologists Bulletin* **40**,159 (1993).
21. P. R. Ehrlich, and H. A. Mooney, *BioScience* **33**, 248 (1983).
22. F. J. Hitzhusen, *Agricultural Ecosystem Environments* **46**, 69 (1993).
23. A. M. N. Izac, K. A. Anaman, and R. Jones, *Agricultural Ecosystem Economies* **30**, 265 (1990).
24. P. M. Vitousek, *Ecology* **75,** 1861 (1994).
25. J. Cairns, Jr., in *Engineering within Ecological Constraints* ed. P. C. Schulze, National Academy Press, Washington, D. C. (1995 b).
26. B. W. Van Wilgen, R. M. Cowling, and C. J. Burgers, *BioScience* **46,** 184 (1996).
27. E. Culotta, *Science* **268**, 1688 (1995).
28. D. H. Janzen, *American Entomologist* Fall : 159 (1991).
29. G. W. Davis, G. F. Midgley, and M. T. Hoffman. *South African Journal of Science* **90**, 319 (1994).
30. A. E. Lugo, *Ecological Applications* **5,** 956 (1995).
31. J. B. Callicott, *Environmental Professional* **13**, 235 (1991).
32. T. M. Power, *Conservation Biology* **5,** 395 (1991).

TROPICAL MOIST FOREST RESTORATION ON AGRICULTURAL LAND IN LATIN AMERICA

KAREN D. HOLL

Environmental Studies Board
University of California
Santa Cruz, CA 95064, USA

1 Introduction

Recent estimates of tropical moist forest land area in Latin America range from 416-687 million ha[1] . Tropical moist forest (TMF) is generally defined by high rainfall (1,800 to 4000 mm annually), even distribution of solar radiation throughout the year, constant high temperatures (mean monthly temperature > 24°C), and lack of frost[1]. Some TMFs receive relatively equal rainfall throughout the year, while others experience a distinct dry season.

Tropical moist forests are being cleared at an alarming rate. While rates of tropical deforestation are difficult to calculate, they are unquestionably high. Myers[2] estimates that approximately half of TMF has been destroyed worldwide; much of this clearing has occurred in Latin America. For example, in Costa Rica, forest cover has dropped from 80 to 25 % in the past 50 years[3]. Skole and Tucker[4] estimate that in the Brazilian Amazon, which covers 4 million km^2, 6 % of forest has been cleared and an additional 9 % of forest exists as small isolated patches or suffers from edge effects. The vast majority of the forest clearing in Latin America is for agricultural purposes, in particular, to create pasture land for cattle grazing[5,6].

Tropical deforestation has profound effects on carbon cycling, soil stability, the hydrological cycle, and the conservation of biodiversity. Forest clearing results in the release of large quantities of CO_2 and trace gases to the atmosphere which will likely affect the global climate[7]. Modelling studies of the Amazon predict a 20 % reduction in rainfall with complete conversion of forest to pasture [8,9]. Tropical forests are home to half or more of all plant and animal species, dispite occupying less than 7 % of the terrestrial surface of the planet[10].

Interest in restoring TMF is increasing in areas being used for agricultural land for a number of reasons. First, in much of the neotropics agricultural land rapidly loses productivity due to poor soils, and is soon abandoned[11,12]. Only about half of tropical land cleared for agriculture is used for this purpose for more than a few years[7]. Second, in some countries government subsidies that encourage forest conversion to agricultural land are being changed[13]. Third, people are increasingly recognizing the importance of the services provided by forest ecosystems, such as controlling erosion, providing food and fuel, and conserving biodiversity[14].

This chapter reviews tropical moist forest restoration in Latin America. Because our understanding of TMF dynamics is minimal, the first section reviews factors limiting TMF recovery. The second section focuses on strategies for facilitating forest recovery, drawing on the results of recent research in a number of Latin American countries. Conclusions drawn from these results must be treated tentatively as most efforts to restore TMF have only been initiated in the past five to ten years. Finally, I briefly discuss socioeconomic obstacles to TMF recovery and restoration.

There is extensive ongoing debate about the definition of ecological restoration. Most scientists agree that it is impossible to recreate an exact replicate of the ecosystem that was destroyed[15]. Therefore, I focus on efforts to restore the functions and at least a subset of the species present prior to disturbance. Increasingly, sustainable agroforestry systems are being developed. These systems serve an important role in balancing human needs and maintaining some ecosystem services. They have been reviewed elsewhere and are only referred to briefly here[16,17,18].

2 Factors Limiting Recovery

One management option for degraded ecosystems is to remove existing stresses, such as grazing, and allow the ecosystem to recover naturally[19]. While slash-and burn agricultural systems have long resulted in abandoned agricultural lands[20], in the past these lands were subjected to less intensive and shorter-term disturbance compared to current practices. Research in the Amazon Basin[21,22] suggests that pastures that have been subjected to light degradation (e.g. cleared but not grazed or extensively seeded with pasture grasses) recover rapidly; however, with increasing length and intensity of disturbance natural recovery is progressively slowed. Similarly, in a study by

Reiners *et al.*.[23] The vegetational composition on two 10-20 year old secondary forest sites that had been cut and not used for pasture was much more similar to primary forest than the vegetational composition of a single site that had been used for pasture before being abandoned. While many tropical forest pioneer species show extremely rapid growth rates[24], a number of factors impede natural recovery in these ecosystems.

2.1 Seed availability

Recovery of tropical moist forest is limited by availability of seeds of non-pasture species. Most tropical forest seeds have an extremely short duration of viability[25]. Accordingly, several studies have shown that seeds of forest species are commonly absent from soil seed banks in pasture land[25,26,27]. In agricultural land that has been used for any length of time, roots of prior vegetation have been destroyed so there is no opportunity for resprouting. Therefore, all seeds of forest plant species must be recently dispersed into agricultural land.

The vast majority of TMF plant species are animal dispersed[28] and many birds and mammals will not venture into disturbed areas. As a result, the few studies of seed rain in Latin American TMF suggest that seed rain declines rapidly within a few meters of the forest edge[29,30,31] and that most seeds falling in pasture land are from species already present in the pasture or from a few small-seeded pioneer species[29,31,32]. Bird activity and seed rain are generally concentrated under remnant trees or shrub patches[27,29,33,34]. In summary, few forest seeds arrive in the pasture and those that do are highly patchily distributed.

Seeds of forest species that do arrive in the pasture are often subjected to high rates of predation[30,32,34,35,36,37,38]. For example, in a seed predation study in Brazil, Nepstad *et al.*[32] reported that all seeds of one tree species placed in the pasture were removed within 24 hours, while more than 80 % of eight of 11 species were removed within 50 days. Similarly, a study in Costa Rica found that approximately two-thirds of all seeds were predated within 30 days[34]. Common seed predators include small mammals, ants and bruchid beetles. Rates of seed predation are quite variable between species which affects patterns of forest recovery[30,32,34].

2.2 Microclimate

A second factor that may limit TMF recovery is harsh microclimatic conditions. Light levels and air and soil temperatures are commonly elevated and humidity and soil moisture levels are reduced in agricultural clearings as compared to forests[39,40]. While exotic pasture grasses are often adapted to high light and low soil moisture levels, stressful microclimatic conditions commonly affect seed germination, and seedling growth and survival of colonizing woody species. Research in Brazil demonstrated that seed germination was low in open pasture as seeds rapidly desiccated[29]. Nepstad *et al.*[12] reported that soil water potential in the top 15 cm of soil was below - 1.5 MPa for a number of weeks during the dry season in the Brazilian Amazon. In related research, xylem pressure potential was two to five times lower for seedlings planted in pasture compared to treefall gaps[27]. Clearly, the extent to which water stress limits TMF recovery is dependent on the degree of rainfall seasonality of the ecosystem.

2.3 Competition with pasture grasses

A third factor that has been repeatedly shown to limit establishment of forest species in agricultural land is the extremely aggressive, exotic pasture grasses that often form a monoculture in previously grazed areas[32,38,41]. Nepsted *et al.*.[32] suggested that grasses limit recovery in the following ways:(1) they do not attract seed dispersers; (2) they provide shelter for seed and seedling predators, such as rodents and leaf-cutter ants; (3) they compete for soil moisture; (4) they increase the probability of fire which kills tree seedlings. Miriti (in press) reported that killing grasses with herbicide increased seed germination in Brazil. Higher cover of non-grass, herbaceous and shrubby second growth species was recorded in areas cleared of pasture grasses in Nicargaua[42] and Costa Rica[31]; however, in neither of these studies were forest plant species observed in cleared areas. In addition, fine root biomass is often much higher in pasture than adjacent forest[31], suggestion elevated below-ground competition in pastures.

In contrast, Aide and Cavelier[30] suggest that grasses may actually facilitate recovery on severely degraded sites by reducing loss of soil moisture. Their research in Columbia showed higher seed germination in plots with grass than those that had been burned or cleared, while seedling growth was similar in all plots. Measurements in southern Costa Rica indicate that light levels and soil temperatures are lower, while humidity and soil moisture are higher, under thick pasture grasses compared to areas of cleared of pasture grasses. Therefore, grasses may provide a more favourable environment for seed germination and seedling survival than open pasture. In summary, grasses may have both positive and negative effects on forest recovery depending on the aggressiveness of the grasses, the severity of disturbance, and the seasonality of the ecosystem.

2.4 Soil physical and chemical characteristics

In some cases, lack of nutrients appears to limit forest recovery in agricultural land. Phosphorous is the most commonly limiting nutrient in tropical forests[43]. However, conclusions about whether or not nutrients limit recovery are highly variable among studies. Results of some reciprocal transplant studies suggest that seedlings are nutrient limited in agricultural land[30,44]. Buschbacher *et al.*.[21] reported that neither vegetational biomass nor vegetational species richness in abandoned pastures was significantly correlated with soil nutrient levels; however, overall nutrient stocks were higher on less degraded land, suggesting that successional vegetation rapidly assimilates excess nutrients. Others research suggests that nutrients do not limit seedling growth.[27,31,32] Uhl[36] suggests that colonizing species may compensate for lowered nutrient levels by allocating more energy to root production and using nutrients efficiently.

Janos[45] suggested that tropical forest recovery may be limited by lack of mycorrhizae, as many colonizing plants are facultatively mycorrhizal while many mature forest species are obligately mycorrhizal. However, there have been few field studies testing this conclusion. Results of a study in Costa Rica suggested that VAM inocula in pastures was high[46]. However, further research is needed in this area.

It has been demonstrated that pasture lands are highly compacted compared to forests[21,23,31]. Soil compaction can affect water infiltration, oxygen and CO_2 diffusion, and root growth. In my research in Costa Rica, infiltration rates in pasture and forest varied by two orders of magnitude; average time for infiltration of 10 cm of water in an abandoned pasture was 193 min (range 43 to 299 min) compared to 1.5 min(range 0.7 to 3.7min) in adjacent forest. However, I am aware of no studies directly investigating the effects of soil compaction on tree seedling growth and survival in the neotropics. Soil compaction did not affect percent germination in a study in Brazil[29].

3. Strategies for Accelerating Recovery

Because tropical forest recovery may be slow in highly degraded areas, humans are increasingly intervening in an effort to accelerate forest recovery. However, few results of restoration projects are available in the scientific literature for a number of reasons. First, most tropical forest restoration efforts have been initiated in the past few years, so long-term data are not available. Second, restoration tends to be a trial and error process and negative results are less likely to be published. Third, many efforts have been initiated by small, grassroots groups that often do not widely disseminate their results. My discussion of TMF restoration draws on the few published studies, as well as my research in southern Costa Rica.

Tropical forests host such a large number of species that reintroducing even a small proportion of the total species would be time and cost prohibitive. Therefore, efforts to restore TMF have generally focused on strategies for facilitating succession rather than trying to either seed or plant a wide variety of species, as is often attempted in temperate systems.

3.1 Planting nurse trees

The most commonly used strategy for accelerating tropical forest succession is planting seedlings of a few native forest tree species that are fast-growing, drought resistant, and able to grow in low soil nutrient levels. The planting of seedlings can result in higher understory diversity[47] and improve soil structure and soil nutrient availability[47,48,49]. Trees also serve to ameliorate extremes in temperature and soil moisture in the understory. Ultimately, the planted saplings will provide the canopy architecture to encourage perching and seed dispersal by birds. Research by Parrotta[50]. suggests that overstory composition resulting from planted seedlings has a large influence on understory colonization rates.

Large areas are commonly planted with seedlings. However, past research suggests that natural colonization is highest at the edge of plantations and naturally occurring woody patches in abandoned pastures[29,50]. These results suggest that the most cost effective method of restoring abandoned pastures is to plant in small patches rather than planting large areas. However, recovery will clearly depend upon the pattern of existing vegetation and the availability of forest patches. More research is needed on the effect of landscape pattern and forest recovery.

Until recently most tropical areas were planted with just three genera: *Pinus, Eucalyptus,* and *Tectona*[51]. Increasingly, projects are being initiated to test a wide variety of native tree species for their ability to grow rapidly in stressful pasture conditions[52,53,54]. Most often, seeds are collected and seedlings are raised in a nursery for three months to one year prior to planting. Trees are rarely directly seeded because of the high losses due to predation and failed germination. Nepstad *et al.*[32] suggest that large, hard-coated seeds may be placed directly on the soil surface because they are less susceptible to predation and desiccation. However, other research suggests that large seed size does not necessarily reduce the probability of being predated[34].

Starting and maintaining seedling in the tropics can be challenging. First, seed collection can be difficult as many TMF trees do not set seed every year[55] and some species must be collected directly from the tree to ensure high generation rates. Second, little is known about the germination requirements for most tropical species. Third, many tropical forest seeds rapidly lose viability when dried, making storage impossible. Fourth, in areas where pasture grasses are dense, it is necessary to clear grasses from around planted seedling with a machete every few month for a year or two to prevent the grasses from completely shading out the seedlings.

Despite the obstacles, a number of studies have shown high seedling survival and growth rates for native species in the first few years[52,53,54]. Many species have more than 80 % survival rates and grow as much as 2.5 m per year.

Survival rates may be largely affected by the specific site condition, such as slope aspect and soil nutrients[52] and by the specific mother tree form which seeds were collected[53]. In seasonal TMF species may not be able to survive extreme microlimatic conditions. Uhl[36] reported that all seedlings planted in a recently abandoned pasture died within one year, while 90 % of seedlings planted under the vegetation of a five-year-old abandoned farm survived for 4.5 years. Similarly, Kolb[29] reported that seedling had higher survival rates in small woody patches in an abandoned pasture than open areas.

Herbivory on young seedlings is often high. For example, in my research in Costa Rica, within the first three months following planting, the stems of 30

% of planted seedlings were cut by rabbits (*Sylvilagus* spp.) and 38 % were excavated by hog-nosed skunks (*Conepatus* spp.). Nepstad *et al.*[32] reported that 30-80% of seedlings of four tree species planted in abandoned pasture were defoliated by leaf-cutter ants (*Atta sexdens*) after 16 days. Wire mesh or plastic tree protectors have been shown to reduce seedling predation in other ecosystems[56]; however, they increase the time and cost of planting efforts. Nepstad *et al.*[32] suggest the possibility of using fungicides on leaf-cutter ant colonies to control ant herbivory, recognizing the potential environmental damage from using chemicals. Carpenter(pers.comm.) has dug up ants colonies to reduce ant herbivory; while effective, this process is extremely time and labour intensive.

Sometimes fertilizers are added to increase nutrient levels when seedlings are planted, which may increase seedling growth. However, using fertilizers substantially increases planting coats and fertilizers are rapidly washed away in areas of heavy rainfall. Moreover, widespread use of fertilizers may inhibit succession by encouraging the establishment and growth of weedy species[36].

3.2 Vegetative propagation

In some tropical regions pieces of tree branches or trunks, referred to as stakes, have been used as "living fence posts" for many years. The stakes are planted and often resprout providing a more permanent structure to which barbed wire is attached. Increasingly, the potential of this practice for restoration is being recognized and the ability of a wide variety of native tree species to resprout from stems is being tested[57]. For example, in northeastern Nicaragua more than 20 species have been tested with mixed success. A second technique, air rooting, has proven successful for the establishment of fifteen species. For air rooting, the cortex of a sapling is cut, treated with growth hormones, and wrapped with soil in plastic. After the sapling have developed roots at the site of the cut, the section is separated and planted.

More research is needed to determine whether these techniques are widely applicable. These techniques remove costs of seed collection, nursery maintenance, and grass clearing, and minimize seedling losses due to herbivory. However, they require the presence of a nearby source forest, which is often not available. Furthermore, considerations must be made of the effect of extensive removal of sapling from existing forest.

3.3 Artificial perching structures and slash piles

Another potential means for facilitating recovery, particularly in pastures lacking remnant trees, is the use of artificial bird perching structures. Artificial structures in open areas have successfully encouraged seed dispersal in temperate systems and would appear to have particular potential for use in tropical forests where the seeds of 50-90% of canopy trees and nearly 100% of shrubs and sub-canopy trees have adaptations for animal dispersal[28]. Increased vertical structure may serve to attract birds farther into the pasture, thereby enhancing seed dispersal.

A few recent studies have investigated the use of bird perching structures in tropical forests[30,31,38,42]. All studies have used posts or branches ranging from 2-5 m height with a perching structure (crossbar or platform) on the top. In two studies, Aide and Cavelier[30], the perching structures were baited with fruit. In all studies, a variety of fruit-eating birds were observed on perches and seed rain under perches was elevated over that in open pasture. Most seeds falling below perches were of pasture or pioneer species. My research shows that while higher than in the open pasture seed rain is still much lower than in the forest. Miriti (in press) and Holl (unpub. data) found that seedling establishment was not elevated under perches as compared to the open, whereas Ferguson[42] reported higher plant species richness below perches, particularly when grasses were cleared. Other studies were too short (<1 yr) to measure differences in plant establishment. It is reported that bird visitation rates and seed rain under baited perches was much higher than non-baited perches. His results also suggest that perch height did not affect perch use. My research in Costa Rica indicates that perches consisting of tree branches are more effective in increasing seed dispersal than posts with crossbars.

Piles of branches and logs remaining after forest felling in areas of slash-and-burn agriculture reduce light levels and temperatures at the soil surface and facilitate woody seedling establishment. They also serve as perching structures and shelter for a number of bird species. Twenty-one percent of shrub patches in a pasture in Costa Rica abandoned for one year were established under the few snags present. Artificially created piles of logs resulted in higher woody seedling establishment during the first year following pasture abandonment in Venezuela; however, the number of seeds of woody species in the soil was not elevated over open pasture[22]. I am not aware of studies investigating the effect of slash piles on long-term forest recovery.

From the limited data available it is impossible to make conclusions about the utility of bird perching structures and slash piles in accelerating recovery. Long-term data is necessary to assess their effectiveness. Perches are relatively cheap to install compared to planting seedlings, but do not have the potential to ameliorate microclimatic conditions or increase soil nutrients. Slash piles may be a more effective way to increase establishment of woody seedlings.

3.4 Clearing existing vegetation

As discussed previously, dense cover of exotic pasture grasses may impede forest recovery. Nepstad *et al.*[32] suggest that burning is essential to rapid forest recovery. Burning serves to reduce grass competition and increase soil nutrient availability. While burning may facilitate establishment of woody plant species, it comprises a large risk to nearby ecosystems and human communities. Even with firebreaks, fires may rapidly spread beyond the desired area.

Grasses may also be killed using herbicides, but hardy grasses often require repeated treatments with pesticides to eradicate all grasses. The use of herbicides is cost-prohibitive on a large scale and may cause extensive pollution of rivers and lakes. In addition, clearing all vegetative cover on compacted soils may greatly increase erosion. Clearly, the costs and benefits of clearing existing vegetation must be weighed carefully.

3.5 Fire prevention

While burning an area at the time of abandonment serves to facilitate recovery, fire after abandonment kills most woody seedlings, thereby setting back the process of succession[22]. Human activities have increased both the frequency and intensity of fires in tropical forests in Latin America, and many tropical forest plant species are not adapted to fire[11,12]. Therefore, an essential component of any tropical forest restoration effort is fire prevention in order to allow for recovery. For example, a great amount of resources are devoted to prevention of the outbreak and spread of fires in northern Costa Rica in a dry forest restoration project[58]. Similarly, in the seasonal wet forest and agricultural land in Costa Rica where I do research, horseback patrols are used during the dry season for fire prevention.

4 Other Obstacles to Restoration

As discussed in the previous sections, there are numerous ecological obstacles to TMF recovery in agricultural land. However, any discussion of TMF restoration would not be complete without mentioning a number of other obstacles to recovery, ranging from cost to human population growth, which, in reality, override any ecological impediments. A detailed discussion of these other factors is beyond the scope of this chapter. However, I mention a number of them.

4.1 Cost and competing land uses

Clearly, the predominant obstacles to most restoration projects are cost and competing land uses. While restored lands may provide long-term income from logging or other extractive activities, short-term income from cash crops, cattle, or subsistence agriculture are lost. For this reason, much of the research has focused on agroforestry systems that integrate multiple land uses in an effort to balance immediate human needs and the maintenance of ecosystem services. Secondary vegetation may provide both subsistence and export crops[59]. However, such management models have limited value in forwarding efforts to develop more sustainable land use policies as long as many Latin American economies are dependent on the export of beef, coffee, bananas, and other cash crops that are predominantly produced by large land-owners[6]. These large and owners are often highly subsidized by the government[1]. For example, Browder[60] estimated that Brazil expended $0.22 in the form of subsidies to produce a quarter-pound hamburger that cost an average of $0.26. Fortunately, in some countries subsidies for highly unsustainable land use are changing. For example, in the past few years the Costa Rican government has passed legislation that reduces logging taxes for land owners who subsequently reforest with native tree species[13]. A wide range of policies to control deforestation and encourage more sustainable land use are discussed by Grainger[1].

In addition to lost income, any efforts to facilitate recovery incur further costs. If seedlings are produced locally, nursery facilities must be constructed and supplies, such as bags, shovels, and fungicides, purchased. Labour costs for collecting seeds and maintaining seedlings must be considered. If seedlings are acquired elsewhere, the cost of purchasing and transporting seedlings is incurred. Furthermore, planting and clearing pasture grasses require extensive labour.

While population growth rates have decreased in most countries in recent years, they still remain well above replacement rate in many countries. Estimated population doubling times are 29 and 37 years in Central and South America, respectively[61]. As long as the population continues to grow rapidly, restoration efforts will increasingly compete with the need for additional land for human settlements and agriculture. I hasten to note that changes in land cover are caused by a complex of factors including not only population size, but also individual consumption levels and patterns and technology[1,62]. Population control and lifestyle changes are an essential component of any restoration project.

4.2 Lack of knowledge and infrastructure for communication

While secondary to cost and competing land uses, restoration efforts in Latin America (and worldwide) are hampered by a lack of information about ecosystem recovery and restoration, as well as a lack of infrastructure for communicating such information. Clearly, substantial indigenous knowledge of natural history exists in many regions. However, the land degradation caused by modern agriculture is far beyond the scope of traditional management systems. In addition, much of this information has not been documented. As has been stated throughout, our understanding of even minimally disturbed tropical ecosystems is limited at best. Most efforts to intervene to facilitate recovery have been very recently initiated and poorly documented.

Existing information is often not available to land owners due to a lack of infrastructure for communication. For example. on the ranch where I work in southern Costa Rica the land-owner has been reforesting with native species for over 25 years. However until recently no local organizations have existed to share this knowledge with others. Fortunately, increasing efforts are being made to make information about native plant species available on a wider scale. For example, in Costa Rica a project funded by USAID is aimed at testing a wide variety of native species for reforestation and making this information available through newsletters and workshops[13]. In an innovative effort, cards with pictures of and information about native tree species were recently distributed in cereal boxes in Costa Rica.

4.3 Ecology and Climate

The ecology and climate of tropical countries serves as a final set of obstacles to restoration efforts. Because endemism is extremely high in most Latin American countries and soil, altitudinal, and climate gradients vary over small distance, the tree species suitable for reforestation may considerably vary over a small spatial scale. Factors limiting recovery may vary over similarly small distances. Therefore, results of studies on restoration strategies may have very localized applications.

The same tropical climate, characterized by extensive rainfall and year round high temperatures, that contributes to rapid recovery from natural disturbance (such as treefalls) creates obstacles for restoration of highly degraded land. For example, transportation of seedlings is difficult because roads are often impassable during the rainy season and high humidity makes stored seeds susceptible to fungal attack.

5. Conclusions

It is clear that a number of factors may limit forest recovery in pasture land. The relative importance of these factors depends on the original ecosystem, the history of disturbance, and the landscape pattern. Lack of forest seeds appears to be the overriding factor in most cases. If few forest seeds are available, then grass competition, water stress, and lack of soil nutrients limiting vegetation survival and growth become moot questions.

Past research has demonstrated a number of potential methods for facilitating succession. The most promising strategy appears to be planting nurse trees. However, as discussed previously, research is desperately needed on the suitability of specific native tree species. In addition, this information must be made widely available.

The rapid rate of forest destruction combined with the incredible complexity of TMF in Latin America make the task of restoring these ecosystems appear overwhelming. But some changes give us reason to be optimistic. First, a number of countries are expanding their efforts to protect and inventory minimally disturbed forest[63]. Second, with more environmental education programs worldwide people are becoming increasingly aware of the critical role tropical forests play in maintaining human well-being. Finally, as a result of this increased awareness, more projects are being initiated on both small and large scales to try to restore the structure and function of these ecosystems and, more often than previously, efforts are being made to widely disseminate the results. In some countries, government incentives are being offered to encourage TMF conservation and restoration efforts.

Acknowledgements

The preparation of this chapter was made possible by funding from the American Philosophical Society, the U.S. Department of Energy Global Change Program, and the U.S. National Science Foundation. I am grateful to M. Loik for helpful comments on a previous draft of this manuscript.

References

1. A. Grainger, *Controlling Tropical Deforestation.* Earthscan Publications Ltd., London (1993).
2. N. Myers, *Deforestation Rates in Tropical Forests and their Cliamatic Implications*. London: Friends of the Earth(1989).
3. C. A.Quesada, *Estrategia de Conservacion para el Desarrollo Sostenible de Costa Rica.* San Jose, Costa Rica: Ministerio de Recursos Naturales, Energia,y Minas.(1990).
4. D. Skole, and C. Tucker, *Science* **260,** 1905 (1993).
5. T. Amelung, and M. Diehl, *Deforestation of Tropical Rainforests; Economic Causes and Impact on Development* Tubingen: Mohr. (1992).
6. P. M. Fearnside. *Ambio.* **2**:537 (1993).
7. R.A. Houghton, in *Handbook of Ecotoxicology,* eds. D. J. Hoffman, B. A. Rattner, G. A. Burton, J. Cairms, Jr. Boca Raton, FL Lewis Publishers.(1995)
8. J. Lean, and D.A.Warrilow, *Nature* **342,** 411 (1989).
9. J. Shukla, and C.A. Nobre, and P. Sellers. *Science* **247,** 1322 (1990).
10. E. O. Wilson, in *Biodiversity* ed. E. O. Wilson Washington, D. C. : National Academy Press USA (1988).
11. C. Uhl, D. Nepstad, R. Buschbacher, K. Clark, B. Kauffman, and S. Subler, in *Alternatives to Deforestation* : *Steps Toward Sustainable Use of the Amazon Rain Forest.* ed. A. B. Anderson Columbia University Press. (1990).

12. D. C. Nepstad, C. Uhl, and E. A. S. Serrao.*Ambio* **20**,248(1991).
13. R. P. Butterfield, and R. F.Fisher. J*ournal of Forestry* june , **37** (1994).
14. K.D.Holl , G.C. Daily and P.R.Ehrlich *Conservation Biology* **9,** 1548 (1995).
15. J. Cairns Jr., *Enviourl Professional* **11**, 152 (1989).
16. J. J. Ewel, *Annual Review of Ecology and Systematics.* **17**,245 (1986).
17. J. B. Alcorn, in *Alternatives to Deforestation, Steps Toward Sustainable Use of the Amazon Rain Forest,* ed. A.B. Anderson, Columbia University Press., New York (1990).
18. F. Montagnini, and F. Sancho. *Ambio* **19,** 386(1990).
19. J., Cairns Jr., *Minerals and the Enviourment* **5**, 32(1983).
20. A. E. Lugo, *Enviourment* **30**, 17,41(1988).
21. R. Buschbacher, C. Uhl, and E. A. S. Serrao, *Journal of Ecology* **76,** 682 (1988).
22. C. Uhl, Buschbacher, and E. A. S. Serrao. *Journal of Ecology* **76,** 663 (1988).
23. W. A. Reiners, F. Bouwman, W. F. J. Parsons, and M. Keller. *Ecological Applications* **4**,363 (1994).
24. J. J. Ewel, *Biotropica* **12**,2 (1980).
25. N. C. Garwood, in *Ecology of Soil Seed Banks* eds. M. A. Leck, V. T. Parker, R. L. Simpson, San Diego Academic Press, Inc. (1989).
26. C. Uhl, K. Clark, H. Clark, and P. Murphy. *Journal of Ecology* **69**, 632 (1981).
27. D. C. Nepstad, C. Uhl, C. A. Pereira, and J. M. Cardoso de Silva, In press.*Oikos.*
28. H.F.Howe, *Biological Conservation* **30**,261 (1984).
29. S. R. Kolbs, Islands of Secondery Vegetation in Degraded Pasture of Brazil, Their Role in Reestablishing Atlantic Coastal Forest. University of Georgia, Ph.D. Dissertation.(1993)
30. T. M. Aide, and J. Cavelier. *Restoration Ecology* **2**, 219(1994).
31. K. D.Holl , In review *Ecological Applications.*
32. D. C. Nepstad, C. Uhl, and Serrao, in *Alternative to Deforestation, Steps Towards Sustainable Use of the Amazon Rain Forest.* ed. A. B. Anderson. Columbia University Press. New York,(1990).
33. S. Guevara, S. E. Purata, and E.Van der Maarel, *Vegetatio* **66**, 77(1986).
34. K. D. Holl and M. E. Lulow, In press *Biotropica.*
35. D. H. Janzen, *Annual Review of Ecology and Systems* **2,** 465(1971).
36. C. Uhl, *Journal of Ecology* **75,** 377 (1987)

37. O. O. Osunkoya, *Australian Journal of Ecology* **19**, 52(1994).
38. M. N. Miriti, *Bulletin of the Ecological Society of America* **76**,186.(1995)
39. R. L. Chazdon, and N. Fetcher, in *Physiological Ecology of Plants of the Wet Tropics*. eds. E.Medina, H.A. Mooney, C.Vazqez-Yanez, The Hague, Dr.W.junk Publishers (1984).
40. N. Fetcher, S. F. Oberbauer and B. R. Strain. *International Journal of Biometeorology*. **29,** 145 (1985).
41. M. R. Guariguata, R. Rheingans and F. Montagnini. *Restoration Ecology* **3**,252 (1995).
42. B. G. Ferguson. Overcoming Barriers to Forest Regeneration in a Degraded Tropical Pasture , an Evaluation of Restoration Techniques. University of Michigan , M. S. Thesis (1995).
43. P. M. Vitousek, and Jr. J. Sanford. *Annual Review of Ecology and Systematics* **17**,137 (1986).
44. I. C. G. Vieira, C. Uhl, and D. Nepstad. *Vegetatio* **115,** 91 (1994).
45. D. H. Janos Biotropica **12**, 56 (1980)
46. C. R. fischer, D. P. Janos, D. A. Perry, R. G. Linderman and P. Sollins, *Biotropica* **26,**369 (1994).
47. J. A. Parrotta, *Agriculture, Ecosystems and Enviourment* **41**,115(1992).
48. R. T. Prinsley, ed, in, *The Role of Tress in Sustainable Agriculture*. Dordrecht, Kluwer Academic Publishers.(1991)
49. F. Montagnini, A. Fanzeres, and S. Guimaraes da Vinha, *Journa of Applied Ecology* **32**, 841 (1995).
50. J. A. Parrotta, *Journl of Vegetation Science* **6**,627(1995).
51. J. Evans, *Plantation Forestry in the Tropics*. Oxford , Clarendon Press (1992).
52. R. Butterfield, In, Proceeding of the XIX IUFRO World Congress, Montreal Canada, Div.1,Vol.2. DD 3-14 (1990).
53. F. L. Carpenter, D. Nichols, M. Rosemeyer, and J. Ketter, Proceeding of the society for Ecological Restoration 1995 Conference. (1995).
54. C. Leopold and A. Fineldey, *Restoration and Management Notes* **13**, 215 (1995).
55. D. H. Janzen, in *Tropical Trees as Living Systems* eds. P.B. Tomlinson, and M.H. Zimerman, Cambridge University Press (1978).
56. D. A. Bainbridge and M. W. Fidelibus, *Restoration and Management Notes* **12,** 86 (1994)
57. T. M. Aide, A. M. Eusse, J. Cavelier, and M. Dupuy, *Restoration and Management Notes* **13,** (1995).

58. D. H. Janzen, In, *Rehabilitating Damaged Ecosystems.* Boca Raton, FL, CRC Press.(1988)
59. J. C. L. Dubois, in, *Alternatives to Deforestation: Steps Toward Sustinable Use of the Amazon Rain Forest.* ed A.B. Anderson Columbia University Press.(1990).
60. J. O. Browder, *Interciencia* **12,** 115 (1988)
61. Population Reference Bureau.World Population Data Sheet . Population Reference Bureau, Inc., Washington, D.C.(1994).
62. P. E. Ehrlich and A. H. Ehrlich, *The Population Explosion.* New York: Simon and Schuster (1990).
63. D.H. Janzen. *Vida Silvestre Neotropical* **4**, 3 (1995).

DAMAGED ECOSYSTEMS : REAL OR PERCEIVED ?

ROBERT H. GRAY

Battelle Pantex
P. O. Box 30020
Amarillo, TX 79177 USA

1 Introduction

Many challenges face those involved with developing coordinated and consistent approaches to restoring damaged ecosystems. An often-ignored first step is to determine whether and to what extent system has been damaged in order to focus efforts where needed, based on the real hazards and risks. Although environmental restoration and cleanup provide many positive benefits, negative features include high costs to taxpayers, uncertainties concerning the technologies to be employed and the risks involved, and the high probability that special interest groups, and activists at large, will never be completely satisfied. Issues concerning the future use of restored sites, whether to protect and preserve their natural features or open them to public exploitation, remain to be resolved. This chapter briefly describes the history of two U.S. Department of Energy sites now undergoing environmental cleanup, reviews long-term environmental and natural resource monitoring efforts and more recent cultural resource characterization activities, summarizes the resulting data, and identifies several unresolved issues concerning the question of real versus perceived ecosystem damage.

2 Site Descriptions and Histories

The U.S. Department of Energy's (DOE's) Hanford Site occupies a land area of about 1,450 km^2 in semiarid southeastern. The Columbia River flows through the Site and forms part of its eastern boundary. This part of the river, the Hanford Reach (about 82 km) is the last unimpounded stretch of the Columbia River in the United States, except for the portion downstream between Bonneville Dam and the river's mouth. Because public access to the

Hanford Site has been restricted and the site has been free from agriculture for over 50 years, it has conserved the habitats of, and now serves as a refuge for, various plants and animals.

Nuclear and non-nuclear industrial and research activities have been conducted at Hanford since 1943. The most significant activities environmentally have been the production of nuclear materials and the chemical processing and waste management associated with the major product, plutonium. By-products have included alpha-, beta-, and gamma-emitting radionuclides and various nonradioactive chemicals in gaseous, liquid, and solid forms. Buildings are confined to a few widely spaced cluster along the Columbia River and in the Site's interior. These clusters are connected by roads, railroads, and electrical transmission lines that together occupy about 6 percent of the land area.

The DOE Pantex Plant occupies a land area of about 65 km^2 in the middle of the semiarid Texas Panhandle. This includes about 24 km^2 of agricultural and rangelands that area leased from Texas Tech University for use as a safety and security zone. The topography is relatively flat, characterized by rolling grassy plains and numerous natural playas (ephemeral lakes). There are over 17,000 playas (ephemeral lakes) on the Texas High Plains, mostly less than 1 km in diameter, that receive water runoff from the surrounding areas.

Pantex Plant was used during World War II (1942-1945) by the U.S. Army for loading conventional ammunition shells and bombs. In 1951, the Atomic Energy Commission began rehabilitating portions of the original Plant and constructing new facilities for nuclear weapons assembly. The primary mission of the Pantex Plant is currently the disassembly of nuclear weapons. In addition to assembly and disassembly, Pantex Plant is also responsible for surveillance, storage, maintenance, modification, repair, and nonexplosive testing of nuclear weapons components; and the manufacture of chemical high explosive (HE) components. Current operations involve short-term handling of encapsulated uranium, plutonium, and tritium, as well as a variety of industrial chemicals. Environmental monitoring has been conducted at Hanford for over 50 years[1] and at Pantex for over 20 years[2] to assess potential impacts to individuals and populations that may be exposed to radionuclides, ionizing radiation, and hazardous chemicals. Environmental media sampled have included air, surface and ground water, foodstuffs (fruits, vegetables, grain, eggs, milk), fish, wildlife, soils, and vegetation. The population status of key fish and wildlife species is also determine at Hanford, and similar efforts are being planned at Pantex Plant. Efforts to preserve and protect archaeological and cultural resources at both sites are more recent[3,4,5].

Based largely on politics and public perception[1], the United States Congress is now being asked to commit billions of tax dollars over several decade to clean up mixed (radioactive/chemical) waste disposal sites at Hanford, Pantex Plant, and elsewhere[6,7,8], even though cleanup and the various options to achieve it have yet to be completely defined. Moreover, extensive monitoring and surveillance efforts, spanning decades of nuclear-materials production, suggest that current risks to the public from environmental contamination are minimal[9,10,11,12,13,14,15]. A review of the monitoring programmemes and summary of the data for both the Hanford Site and Pantex Plant follow.

3 Radiological Monitoring

3.1 Air

At Hanford, air is sampled continuously for airborne particulates and analyzed for radionuclides at 39 onsite and offsite locations, including 9 in nearby and distant communities[16]. At selected locations, gases and vapours are also collected and analyzed. The potential radiological dose from various point-source emissions to air to the maximally exposed individual (a hypothetical person who receive the worst case dose) was 0.01 mrem/yr (effective dose equivalent) in 1994, well below both the DOE radiation limit of 100 mrem/yr to the public and the 10 mrem/yr limit for the air pathway specified by the Clean Air Act. At Pantex Plant, air is sampled continuously for airborne particulates and gaseous tritium at 17 onsite and offsite locations[14]. Seventeen additional stations were established around the Plant perimeter during 1994 and 1995[15]. Samples are analyzed for radionuclides (^{3}H, ^{234,238}U, $^{239/240}$Pu). The potential radiological dose to the maximally exposed individual was 0.000058 mrem in 1994, well below the average external background does for the Texas Panhandle (110 mrem/yr), the DOE radiation limit of 100 mrem/yr to the public, and the 10 mrem/yr limit for the air pathway specified by the Clean Air Act.

3.2 Surface Water

Columbia River water is used for drinking at cities downstream and for crop irrigation and recreational activities (fishing, hunting, boating, water-skiing, swimming). Thus, it constitutes a potential environmental pathway for radioactivity in liquid effluents to reach people. Although radionuclides occur in small but measurable quantities in the Columbia River, concentrations are below Washington State and U.S. Environmental Protection Agency (EPA) drinking water standards.

Pantex Plant does not include or border on any rivers or streams. However, storm-water runoff from the Plant and lands leased from Texas Tech University flows through ditches to onsite playas and in sheetflow offsite. Thus, playas are ideal surface locations for assessing release from the Plant. Radionuclide concentrations in 1994 were less than DOE's Derived Concentration Guide (DCG) for ingested water and the levels established in the Safe Drinking Water Act (SDWA). Gross alpha/beta levels observed in 1994 were comparable to historical and offsite control levels.

3.3 Ground Water

At Hanford, ground water, primarily from an unconfined aquifer, is currently sampled from over 800 wells and analyzed[16]. Tritium (^{3}H), which occurs at relatively high concentrations in the unconfined aquifer, is one of the most mobile radionuclides, and thus its distribution reflects the extent of ground-water contamination from onsite operations. Ground water from the unconfined aquifer enters the Columbia River through subsurface flow and springs that emanate from the riverbank[17]. Tritium concentrations in wells near the springs ranged from 19,000 to 250,000 pCi/L and averaged 176,000 pCi/L in 1985[18]. Although the concentrations of ^{3}H and other radionuclides in springs generally reflected those in nearby ground-water wells, they were lower in the springs due to the mixing of ground and surface water. Tritium concentrations in the river were generally less than those in springs and were less than half of the regulatory limit for drinking water (20,000 pCi/L)[19]. From 1983 through 1989, annual average ^{3}H concentrations in the river (< 200 pCi/L) were at least a factor of 100 below the drinking water limit[10]. Other radiological constituents found above detection limits in the river have been ^{99}Tc, and $^{234,238}U$ and ^{90}Sr, ^{129}I, ^{235}U and $^{239,240}Pu$[10,20].

At Pantex Plant, ground water has been sampled from 19 onsite wells (nine in the Ogallala aquifer, and ten in overlying perched ground water)[15]. Eight onsite drinking water taps were also sampled in 1994. One offsite location each was sampled for ground water and drinking water as controls. Naturally occurring radionuclides were present in the Ogallala aquifer in 1994, and concentrations were less than the DOE DCG and the levels established in the SDWA. Although contamination (e.g. HE) has been found in perched ground water beneath the Plant and to some extent offsite to the southeast, there is no indication of contamination from Pantex activities in the Ogallala water supply wells or a well field operated by the City of Amarillo just northeast and downgradient of the Plant.

3.4 Foodstuffs

Samples of alfalfa and several foodstuffs, including milk, vegetables, fruit, beef, chickens, eggs, and wheat, are collected from several locations, primarily downgradient (i.e. south and east) of the Hanford Site[10,13,18]. Samples are also collected from upgradient and somewhat distant locations to obtain information on radiation levels attributable to worldwide fallout. Foodstuffs from the Riverview Area (across the river and southeast of the Site) are irrigated with Columbia River withdrawn downstream of the Site. Although low levels of ^{3}H' ^{90}Sr, ^{129}I, and ^{134}Cs have been found in some foodstuffs, concentrations in samples collected near Hanford are similar to those in samples collected away from and upgradient of the Site. Although human foodstuffs have not been routinely sampled at Pantex, analysis of winter wheat and sorghum that are eaten by cattle showed radionuclide concentrations similar to those expected to occur naturally.

3.5 Fish and Wildlife

Fish are collected at various locations along the Hanford Reach of the Columbia River, and boneless fillets are analyzed for ^{60}Co, ^{90}Sr, and ^{137}Cs. The carcasses are analyzed to estimate ^{90}Sr in bone. Short-lived radionuclides, including the biologically important ^{32}P and ^{65}Zn, have essentially disappeared from the river[21] through radioactive decay. Radionuclide concentrations in fish collected from the Hanford Reach of the Columbia River are similar to those in fish from upstream locations.

Deer (*Odocoileus* sp.), ring-necked pheasants (*Phasianus colchicus*), mallard ducks (*Anas platyrhynchus*), Nuttall cottontail rabbits (*Sylvilagus nuttallii*), and black-tailed jack (*Lepus californicus*) are collected at Hanford, and tissues are analyzed for ^{60}Co and ^{137}Cs (muscle), $^{239,240}Pu$ (liver), and ^{90}Sr (bone). The doses that could be received by consuming wildlife at the maximum radionuclide concentrations measured in 1985 through 1994 were below applicable DOE standards[10,13,18]. Routine wildlife sampling (prairie dogs) was initiated at Pantex Plant in 1995.

3.6 Soils and Vegetation

At Hanford, samples of surface soil and rangeland vegetation (sagebrush) are currently collected at 39 onsite and 14 offsite locations[16]. Sampling and analyses in 1985 through 1994 showed no radionuclide build-up offsite that could be attributed to Hanford operations[10,13,18].

At Pantex Plant, samples of surface soil were collected at 31 onsite and 17 offsite locations during 1994. Plutonium levels were comparable to historical levels and to those at control sites. Of the radionuclides present in soils, ^{238}U appears to be the most prevalent. However, even the highest ^{234}U and ^{238}U levels were comparable to the historical averages established by Honea and Gabocey[22]. All other areas showed levels well below historical averages. Vegetation samples (native, exotic, and domestic grasses) were collected from 5 onsite and 17 offsite locations during 1994. No indication of radionuclide contamination from Pantex Plant was found.

3.7 Overall Radiological Impacts from Hanford and Pantex Operations

The maximally exposed individual is a hypothetical person who would receive the maximum calculated radiation dose when worst-case assumptions are used concerning location, inhalation of radioactive emissions, consumption of contaminated food and water, and direct exposure to contaminants. Expressed as effective dose equivalents, the calculated does for a maximally exposed individual at Hanford was 0.02 to 0.1 mrem annually from 1985 through 1994. The average per capita effective dose for 1985 through 1994, based on the human population of 330,000 to 380,000 people living within 80 km of the Site, was < 0.01 to 0.05 mrem annually[10,13,18]. The calculated dose for a maximally exposed individual at Pantex Plant was ≤ 0.000058 mrem in 1993 through 1994.

The estimate and the measured background radiation levels at Hanford and Pantex Plant can be compared to those from the other routinely encountered radiation sources, such as natural terrestrial and cosmic radiation, medical treatments including x-rays, natural internal body radioactivity, worldwide fall out, and radiation from consumer products (Figure 1). Radiation doses to the public from the Hanford Site and Pantex Plant have been consistently below applicable standards and substantially less than doses from routinely encountered sources of radiation not associated with either facility.

4 Chemical Monitoring

4.1 Air

At one time, nitrogen oxides (NOx) were released at Hanford from fossil-fueled steam and chemical processing facilities, most notably the Plutonium Uranium Extraction (PUREX) Plant. Nitrogen dioxide samples were collected until PUREX Plant ceased operation in 1990. Nitrogen dioxide concentrations measured in 1984-1990 were well below federal and state ambient air quality standards[11,12].

Because Pantex Plant is not a major source of air emissions, only limited nonradiological air monitoring is required. The Texas Natural Resource Conservation Commission (TNRCC) monitors for volatile organic compounds (VOCs), particulates, and fluorides at five onsite and one offsite location. No significant contaminant levels have been reported. Three additional air sampling stations were constructed around the Plant perimeter in 1995 and sampling there will become part of the Plant's routine monitoring programme for nonradiological constituents in 1996.

4.2 Surface Water

Nonradioactive waste water from the Hanford Site is discharged at eight locations along the Columbia River. Discharges consist of backwash from water intake screens, cooling water, water storage tank overflow, a building drain, and fish-laboratory waste water. Effluents from each outfall are monitored under a National Pollutant Discharge Elimination System permit. The Columbia River is also monitored by the U.S. Geological Survey (USGS) upstream and downstream of the Site, to verify compliance with Washington State water-quality requirements.

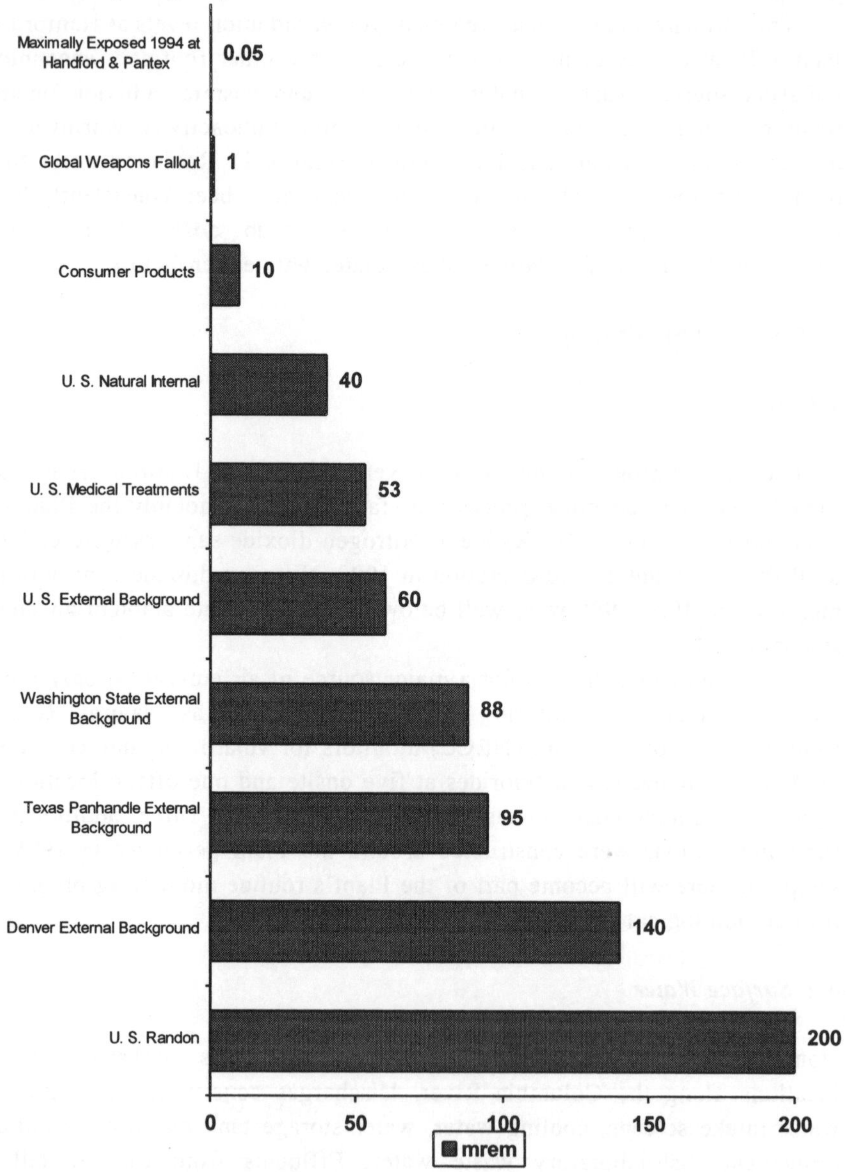

Figure 1 : Average annual radiation doses (per capita) from various sources.

The Columbia River also been sampled in transects both upstream and downstream of the Site[23]. The samples were analyzed for VOCs, metals, and anions based on their known presence in ground water near the river and studies documenting their entry into the river. Although VOCs were not routinely detected, several metals and some anions were detected at levels comparable to those reported by the USGS as part of their national water quality monitoring programme[24]. There were no measurable differences between upstream and downstream concentrations of metals and anions in river water. Becker and Gray[25] evaluated several water quality variables in the Columbia River upstream and downstream of Hanford reactors, for the periods 1951 to 1953 and 1986 to 1988. Five reactors were operational during the 1951 to 1953 period, and all were shut down during the 1986 to 1988 period. Despite the operational differences, levels of ammonia nitrogen, pH, and coliform bacteria in river water did not differ significantly between the two periods. Phosphate in the Hanford Reach had decreased significantly over the 33-years interval, while biochemical oxygen demand, dissolved oxygen, and nitrate nitrogen had increased. These changes, while detectable statistically, were relatively small and not related to current activities on the Site. Today, the quality of water in the Hanford Reach remains well within Washington State standards for Class A waters[26]. Low pH values, that appear to originate upriver occasionally violate these standards.

At Pantex Plant, onsite and offsite surface water locations are sampled, and the samples are analyzed for VOCs, semivolatile organics, HEs, herbicides, pesticides, polychlorinated biphenyls, metals, nutrients, and other components reflecting water quality. Results are compared to TNRCC Maximum Contaminant Levels (MCLs). When no MCL is available, results are compared to values previously observed at Pantex Plant, which serve as a baseline. Comparisons to these historic values are primarily for metals, HEs, and inorganic constituents. None of the historic values has ever been a source of concern. Cosampling by the TNRCC confirms this conclusion. In addition, Pantex Plant has been issued a waste-water permit from the TNRCC that requires monitoring of effluents and outfalls. Other than occasional high-pH excursions due to biological activities (e.g. photosynthesis), Pantex Plant has been in compliance with all requirements.

4.3 Ground Water

At both Hanford Site and Pantex Plant, samples are collected from groundwater wells and analyzied for chemical constituents. In addition, onsite drinking water sources are sampled and analyzed for water quality.

At Hanford, detected constituents include several metals, anions, coliform bacteria, and total organic carbon. Many of these constituents are common in natural ground water. However, several hazardous chemical have also been identified in Hanford ground water at concentrations exceeding both existing and proposed federal drinking water standards[13]. These include nitrate, chromium, cyanide, fluoride, carbon tetrachloride, trichloroethylene, and tetrachloroethylene. Wells containing these constituents are mainly near operating areas and are not used for drinking water.

With few exceptions, chemical constituents in drinking water wells at Pantex Plant are well below the MCLs, and there is no evidence of contamination by HE, organic compounds, or metals in the Pantex water supply wells or in the nearby well field operated by the City of Amarillo. The MCL for iron has been exceeded at two wells, but these high values may reflect corrosion of casing material. Water from wells constructed with polyvinyl chloride casings did not exceed the MCL. The occurrence of chormium, which has been detected in drinking water samples from the Pantex Plant and in some control samples, but at levels below the MCL, may also reflect corrosion in pumps and piping[14].

The contaminants chromium, nickel, trichloroethylene, dichloroethane, benzene, and HEs and their degradation product trinitrobenzene have been found in the perched aquifer beneath the Plant. The extent of contamination is being investigated via a Resource Conservation and Recovery Act Facility Investigation, and remediation has recently been initiated.

5 Biological Monitoring

Restricted land use at Hanford has favored native wildlife that frequent riverine habitats, for example, mule deer (*Odocoileus hemionus*), Canada geese (*Branta canadensis*), and great blue heron (*Ardea herodias*). The Site also serves as a refuge for other migratory waterfowl, elk (*Cervus elaphus*), coyote (*Canis latrans*), and a variety of other animals and plants[27,28,29] .

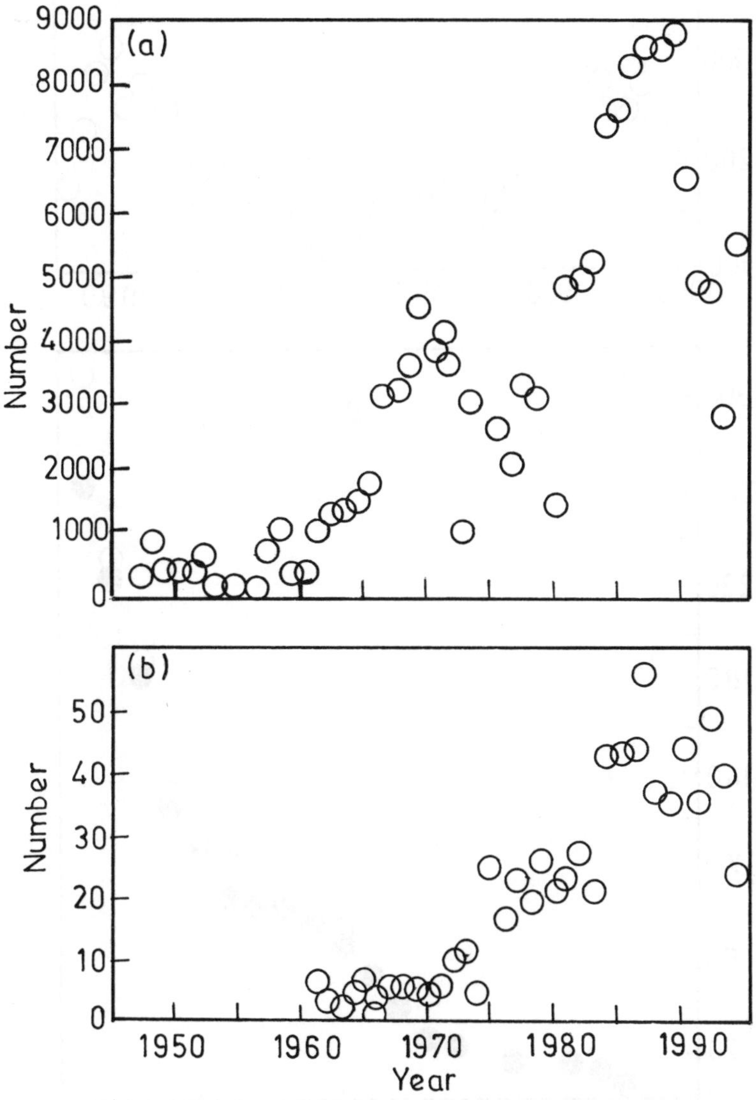

Figure 2 : Number of (a) chinook salmon redds (nests), (b) wintering bald eagles at Handford.

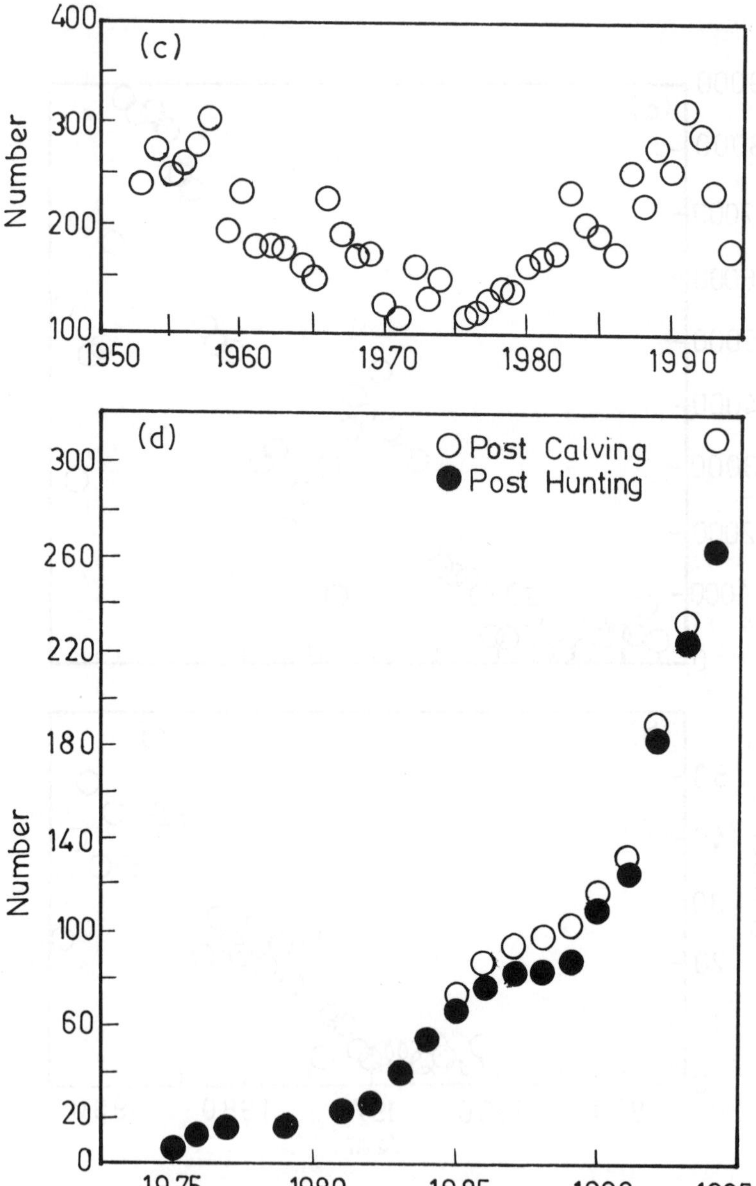

Figure 2 : Number of (c) Canada goose nests, and (d) elk at Hanford.

The Columbia River at Hanford supports 48 species of fish[30] and serves as a migration route for upriver runs of Chinook (*Oncorhynchus tshawytscha*), coho (*O. kisutch*), and sockeye (*O. nerka*) salmon and steelhead trout (*O.mykiss*). Steelhead trout spawn in the Hanford Reach[31], and the Reach supports the last remaining mainstream spawning habitat for all Chinook salmon. Based on redd (nest) counts from the air, fall Chinook salmon spawning in the Hanford Reach of the mainstream Columbia River increased dramatically from 1980 (Figure 2a). Underwater observations by divers[32] showed salmon redds at depths below those visible by boat or aircraft suggesting that salmon spawning in the Hanford Reach may be greater than can be estimated by aerial counting. The increase in salmon spawning attracted increasing numbers of wintering bald eagles (*Haliaeetus leucocephalus*), and the eagle population now fluctuates with the salmon population (Figure 2b). The marked decrease in salmon redds observed in the 1990s is of considerable interest because of the need to preserve salmon runs in the Columbia River watershed.

The sparsely vegetated islands in the Columbia River have historically been used as nesting habitat by Canada geese[33,34]. From the mid-1950s to the mid-1970s, the number of goose nests declined from about 300 to about 100 annually (Figure 2c). From the late 1070s to 1992, the numbering of nests gradually increased to over 300. Initially, closure of the Hanford Reach was beneficial to the geese by providing freedom from human incursions. However, the coyote also benefited and is believed to have caused the decline in numbers of goose nests into the mid-1970s. The Hanford Reach is now reopened for public use and selective coyote control has been practiced near the city of Richland, which may help explain the increase in numbers of nesting geese in the 1990s.

When monitoring began, there were no nesting great blue heron on the Hanford Site. However, there are now four active colonies consisting of 35-40 or more birds each, and herons are present year-round[27]. Their major food source is Columbia River fish.

Elk first arrived on the Hanford Site in 1972[35]. From a small founding population, the herd size grew (Figure 2d) to about 80 animals in 1987[27] and now exceeds 300. The rapid increase in elk is attributed to the lack of predation or human disturbance during calving, absence of onsite hunting, and the lack of competition from sheep and cattle for available forage. The mule deer population at Hanford is estimated at several animals and appears stable even in the absence of onsite hunting. Coyote predation on fawns is believed to be an important factor in maintaining the stable deer population[36].

The Southern High Plains in which Pantex Plant lies include short-grass prairie and farmland, and offer only slight relief expressed as playa basins. In this semiarid region, the playas are typically ephemeral, yet highly productive. Most of Pantex Plant's nonindustrial areas are under cultivation or used for grazing by Texas Tech University or their leasing farmers. Population monitoring and biological studies are now being planned for implementation at the Pantex Plant. To date, some characterization efforts have been completed and several species lists have been developed[37,38,39,40,41].

Agricultural and industrial activities have reduced biological diversity across most of the Southern High Plains, and the playa areas at Pantex Plant have been intensively grazed in the past. The only remaining areas of native habitat at the Plant are small and center on the playas, as aquatic habitat and ephemeral wetlands, and on short-grass prairie upland in and around all playas. Plant operations in several industrial areas, such as the Burning Ground and firing sites, have precluded agricultural activities and therefore have provided some additional protection for native habitat.

During field survey at Pantex Plant in May, July, and September 1993, and June, and September 1995, 262 floral species were identified[38,41]. Because the surveys focused on natural areas of the Plant, playa-related species were strongly represented. Nevertheless, many of the species found were non-native, weedy species, and many native species were represented by only a few individuals.

Although no federal- or state-listed endangered species were found, several species of barrel cactus were noted, a variety of which is protected. The green-flowered barrel cactus *Echinocereus viridiflorus* var. *Viridiflorus* is the rarest species at Pantex Plant, represented by a single population at Playa 1. Another barrel cactus, the more common cerise-flowered *Coryphantha vivipara*, grows on the slopes and short-grass prairie uplands at the playas, the Burning Ground, and the firing sites.

Several animal surveys were conducted in 1993 and 1994. A general survey[39] lists a total of 65 species, including protistans, invertebrates, amphibians, reptiles, birds, and mammals. Field surveys to identify amphibians and reptiles showed few amphibian species, in part because of the lack of precipitation during the summer of 1994.

Seyffert[40] compiled a bird checklist for Pantex Plant that indicates abundances expected during various seasons of the year, based on habitat types and knowledge of bird migrations through the local area. The largest concentrations of waterfowl occur at Playa 1, the only playa onplant containing water year-round. Hundreds of mallard (*Anas platyrhynchos*) and widgeons (*A. american*) and lesser numbers of Canada geese winter on the water and in the cattail cover. Playa 1 is used seasonally by many other species, the most notable being the bald eagle, red-tailed hawk (*Buteo jamaciensis*), northern harrier (*Circus cyaneus*), white-faced ibis (*Plegadis chihi*), black-crowned night heron (*Nycticorax nycticorax*), yellow-crowned night heron (*Nyctanassa violacea*), eared grebe (*Podiceps nigricollis*), canvasback duck (*Aythya valisineria*), and Wilson's phalarope (*Phalaropus tricolor*).

Depending on precipitation and quality of habitat, different playas provide different habitat to various species. During the survey period, Playas 2,3, and 4, and Pantex Lake supported greater numbers of small rodent species than Playa 1. Birds of prey are common in playas that contain water only seasonally. Although white-faced ibises are occasionally seen at Playa 1, they are more common during spring and fall migrations at Playa 2 and 4. Black terns (*Chlodonias niger*) have been seen only at Playa 4.

The short-grass prairie uplands surrounding Playa 2,3,4, and 5 support prairie dogs (*Cynomys ludovicianus*) and form part of the wintering habitat for bald eagle, golden eagle (*Aquila chrysaetos*), ferruginous hawk (*Buteo regalis*), and rough-legged hawk (*B. lagopus*). In summer, prairie dog towns provide nesting sites for burrowing owls (*Athene cunicularia*). Swift foxes (*Vulpes velox*) are known to den near prairie dog towns and there are recent indications that swift foxes may be present at Pantex Plant although none have been seen since 1970. Migrating mountain plovers (*Charadrius montanus*) and other grassland birds also use prairie dog towns as seasonal habitats. A few Texas horned lizards (*Phrynosoma cornutum*), a state-listed threatened species, are also found in short-grass prairie uplands surrounding the playas and at the firing site west of Zone 4.

6 Cultural Resources

Hanford was once populated by Indian tribes and settlers who left an abundance of virtually undisturbed cultural artifacts and archeological deposits[3,5]. The site lies on land ceded to the United State in 1855 under treaties with the Yakima Indian Nation and the Confederated Tribes of the Umatilla Reservation. Early weapons and utensils, hunting trails, and the remains of lodgings and campsites give insight into the thousands of years of human history in the Northwest. This cultural legacy is especially valuable because other vestiges of the past have been lost to hydroelectric development, farming, and industrial construction.

As landlord of the Site, DOE has responsibility to prevent tampering with, or disturbance, removal, and collection of archeological resources by unauthorised individuals. The Archeological Resource Protection Act provides severe criminal and civil penalties for violators. A systematic process of identifying, preserving, and protecting these resources was implemented at Hanford in 1987. Currently, under a cultural resources management plan [3], in-depth review is required before, and often during, any construction work. Constructive sites are visually inspected and historical records are checked. If needed, test excavations are performed and construction activities monitored to identify, evaluate, and assess any potential effects on archeological and historical resources. If items or areas of archeological, historical, or cultural significance are found, they are considered for nomination to the National Register of Historic Places. If eligibility is established, alternatives to the proposed construction, such as moving to another location or using techniques that would lessen impacts, are evaluated and implemented if possible. Most archeological and cultural sites at Hanford are located along the Columbia River and most surveys have been conducted along the shoreline.

Archeological surveys have also been conducted at the Pantex Plant[4], and historic properties have been catalogued[42]. Through these surveys, prehistoric archeological sites have been identified and recorded near playas, but not in interplaya areas. Management procedures protecting these sites are being implemented, including establishment of buffer zones around each playa and prior evaluation of proposed ground-disturbing activities.

7 The Dilemma : Addressing Reality Or Perception ?

A half century of operations at Hanford and Pantex Plant, coupled with today's political and regulatory climate, have created a need for environmental cleanup and restoration. Nationally, Comprehensive Environmental Response and Liability Act cleanup efforts have dealt mostly with one or two contaminants spread over limited areas[43]. Operation of large industrial complexes that generate and process nuclear materials has created an entirely different situation-the need to clean up multiple chemical and radioactive contaminants in several physical environments spread over many square miles. Site cleanup, or site assessments to demonstrate that an unacceptable risk does not currently and will probably never exist, requires enormous planning, financial, and resource commitments spanning decades.

Initial construction and operations at Hanford and Pantex Plant were carried out in secrecy under military oversight, with a single onsite contractor to meet one important need, the defense of a nation. The decision process governing cleanup 50 years later is more complex, and involves as participants various federal and state agencies, multiple contractors, local governments, and the public. At Hanford alone, cleanup has been scheduled to take place over 30 years, with an estimated cost exceeding $1 billion annually[44]. While the public is concerned about having the waste cleaned up, it has so far expressed little concern about the costs, costs that will be paid by current and future generations. Eventually, these costs will have to be compared to the environmental benefits. Thus coordinated and consistent approaches to remediation studies are essential. At Hanford, a Tri-Party Agreement involving DOE, the State of Washington, and EPA evolved as the first step in the coordination process.

Ironically, many cleanup options have the potential to increase rather than decrease exposures, first, directly to workers involved in the cleanup process and, second, indirectly to the public and environment as buried or contaminated waste is temporarily exposed to the elements[45]. Although cleanup may reduce the volume of highly contaminated material, the volume of material with low-level contamination (filters, piping, clothing, etc.) will increase. Materials that become contaminated during the cleanup process will have to be treated, buried onsite, or transported and rebuired elsewhere. If old or newly contaminated material must be treated or reburied elsewhere, the risk of transportation must also be considered. *In situ* cleanup options, such as vitrification (solidifying or glassifying through use of an electric current) and grouting (transformation to cement) are potential treatment methods, although much additional study of these fledgling technology is needed, and grouting has lost favor at Hanford. A pump-and-treat technologies is currently being evaluated at Pantex Plant as an option for cleaning up the perched aquifer beneath the Plant.

Although the decisions to clean up Hanford and Pantex Plant are based on regulatory compliance requirements, it appears that they are also driven by public anxiety and politics, rather than by quantitative risk and cost-benefit analyses based on sound scientific data. The lack of congruence among actual exposures (usually zero), potential exposures, and environmental regulations, coupled with high cost, has caused both scientists and the public to question the wisdom of some cleanup initiatives. The possibility of exposure now or in the future is so low at some sites that knowledgeable critics are asking : Why should expensive cleanup and remediation efforts even be undertaken ?

As discussed above, environmental surveillance and monitoring have been conducted at both Hanford and Pantex Plant for decades to ensure that public exposures to radiation and chemicals were maintained as low as reasonably achievable and to assess potential environmental impacts. Resulting data show that the current environmental and public health risks are low[2,9,10,11,12,14,15,46].

The current radiological doses (Figure 1) at Hanford and Pantex Plant are orders of magnitudes less than the exposure received each year from natural background radiation, or while traveling on a coast-to-coast airline flight, or while simply residing in your home. Furthermore, the small amounts of chemical and radioactivity now entering the Columbia River via Hanford site groundwater are but a fraction of those released from reactors in cooling water discharges without damage to the public or environment, including directly exposed aquatic biota[25,29,47,48].

The highest radiological doses to people from past Hanford released to the air occurred from 1944 to 1951. Iodine-131, a short-lived radionuclide, accounted for more than 98% of the dose. Other radionuclides such as plutonium, ruthenium, strontium and cerium accounted for less than two percent of the dose. People could have been exposed to airborne radionuclides through inhalation, consumption of food crops, and drinking fresh milk from cow pastured downwind of Hanford. Actual radiological doses depend on location, life-style, age, and other factors. Dose estimates suggest that the most exposed group included infants and small children who drank milk from cows pastured downwind of Hanford particularly during 1945[49].

The highest doses to people from past radiological releases to the Columbia River occurred from 1956 to 1965. Five radionuclides, phosphorus-32, zinc-65, arsenic-76, sodium-24, and neptunium-239 accounted for 94% of the river dose. People could have been exposed to waterborne radionuclides directly through swimming, drinking river water, or eating Columbia River fish. Again, radiological doses depend on location, life-style, and other factors. Annual doses to most individuals were less than 5 mrem. Only those individuals who ingested large quantities of resident fish could have received annual doses in excess of 140 mrem[50].

Because public access to the Hanford Site has been restricted and the Site has remained free from agricultural use for 50 years, it has preserved the habitats of, and now serves as a refuge for, a variety of native plants and animals (Figure 2), some of which have recreational and commercial value[27,28,29]. An extensive series of assessment studies have revealed no evidence of significant effects on anadromous fish resulting from Hanford operations[29,47,48,51]. Rather, the number of salmon redds (nests) constructed in the unimpounded Hanford Reach increased(Figure 2a) as other spawning areas were inundated behind main-stem dams on the Columbia River. Concomitant with increased salmon spawning, the number of bald eagles wintering on the Hanford Site increased (Figure 2b). An elk herd, established by immigration of a few individuals in 1972, also increased under the protection of the Hanford Site. In addition, the Site serves as a nesting area and refuge for such other fauna as Canada goose and great blue heron, and it supports stable or growing populations of mule deer, coyote, and other animals and various plants[27,28,29,34].

Because of its uniqueness, part of the Site was established as the Arid Lands Ecology Reserve, on the northeastern slope of Rattlesnake Mountain, in 1967 to provide an outdoor laboratory for environmental research in an undisturbed shrub-steppe ecosystem. In 1977, the entire site was dedicated as a National Environmental Research Park. In 1992, DOE signed a cooperative agreement with The Nature Conservancy to characterize, preserve, and protect rare biotic species and communities at Hanford. These activities could not be conducted elsewhere because of human intrusion.

What will happen to the Hanford Site once the disposal areas for radioactive and other materials have been cleaned up? If the site is returned to public use for agriculture, as occurs over much of eastern Washington, the resulting destruction of natural habitat would severely impact Hanford's wildlife resources and destroy its value as an ecological research area. Agricultural runoff would lower water quality in the Columbia River and perhaps damage anadromous fish runs and the resident fishery. Given that current public and environmental risks from Hanford are minimal, and given Hanford's current value as an ecological refuge, why clean it up at all[44,52] ?

In the past, some waste streams containing low-level radioactive and chemical wastes were discharged to the ground. In addition, 149 single-shell and 28 double-shell storage tanks located underground in the center of the Hanford Site contain high-level radioactive waste and chemicals. Tank capacities range from 55,000 to 1,000,000 gallons. Some of the older, single-shell tanks are known to have leaked; 66 single-shell tanks are assumed to leak. The total volume of radioactive liquids estimated to have leaked from the tanks over the years exceeds 750,000 gallons. While some scientists believe the bulk of the leaked material has been absorbed in the upper layers of substrate, some mobile contaminants from decades of surface discharges and possibly tank leaks have reached ground-water beneath the site and are moving with ground-water flow to the Columbia River. Although travel times are slow (measured in decades) and concentrations decline with time and dilution, there is no way to guarantee that some contaminants will not eventually enter the river at levels that might be of environmental concern. This possibility has resulted in great sociopolitical debate. Furthermore, there is no guarantee that additional tanks will not eventually leak. Although no health impacts have been demonstrated to date, prudent practice dictates that basic studies to identify viable cleanup options be conducted, that quantitative risk and cost-benefit analyses be performed, and that investments be made where they count most.

In 1942, best judgments were made quickly to select a suitable site for the nation's first nuclear production facility, to determine the physical separation required between and among reactors and reprocessing plants, and to identify the best solution for disposal of waste streams. For the most part, these judgments were sound and have withstood the test of time.

Today, best judgments based on cleanup goals, technology capabilities, risk-reduction benefits, public acceptance, and costs will govern the selection of cleanup strategies and aid in determining Hanford's future land uses. Intended future land use will be important in selecting cleanup goals. Options that involve agricultural or other public use will require stricter cleanup goals than will those for industrial uses. The Hanford Future Site Use Working Group was established in 1992 to actively solicit public input to potential future land use options. The group has representatives from DOE; EPA; Department of the Interior; the states of Washington and Oregon; Benton; Franklin; and Grant counties; Native Americans; agricultural and business interests; environmental and activist groups; and the cities of Richland and Pasco.

Another action that may affect future land use along the Columbia River was initiated by Congress (Public Law 100-605) in 1988. That action required a comprehensive study of the Hanford Reach. The Secretary of the Interior, in consultation with the Secretary of Energy, was to (1) inventory and evaluate the river's resources, and (2) develop and analyze a series of protection alternatives, including designation of the Reach in the National Wild and Scenic Rivers system. During the eight-year study period, which started in 1988, no federal agency could build dams or conduct navigation or channelization activities such as dredging within the Reach. All other activities had to be planned and implemented to minimize adverse impacts on the river's resources.

The National Park Service (NPS) is the lead agency for the Department of the Interior and has announced its preferred alternative, establishing a National Wildlife Refuge with a Wild and Scenic River designation, in a draft Environment Impact Statement (EIS)[53]. The final report to Congress will incorporate public comments and present the study team's recommendations. Alternatives considered in the draft EIS, in addition to the preferred alternative, include designation of a National Conservation Area or a National River, prohibiting dams and dredging in the Hanford Reach with no additional designation, or taking no action. If the Hanford Reach is declared a federal Wild and Scenic River, the NPS would be required to preserve the river in its natural state and allow its resources to be used not changed, altered, or depleted. Private landowners holding river frontage along the Hanford Reach are already protesting this possibility.

To date, no one has asked U.S. taxpayers if they were willing to pay the price of over one billion dollars annually, to support Hanford cleanup efforts. Local governments favor cleanup because it creates jobs, and Northwest politicians favor cleanup because it is politically expedient. However, the response of taxpayers throughout the United States will be interesting, given that the national debt currently exceeds three trillion dollars and annual deficit spending is the rule rather than the exception. The DOE's cleanup budget has already been drastically reduced, resulting in the loss of thousands of jobs at Hanford alone. The outcome of the budget debate between Congress and the President of the United States that resulted in a partial government shutdown in late 1995 and early 1996 will likely have implications to future cleanup efforts.

Regardless of which cleanup options are chosen or the level of cleanup to be attained, environmental monitoring will continue as an aide to waste management during the cleanup processes and for decades beyond. Monitoring is required to assure worker, public and environmental safety. Current modifications to monitoring programmes call for increased emphasis on sampling and analysis for hazardous chemicals and on monitoring those areas where the public can come in contact with the sites. At Hanford these include the Columbia River, its shoreline and islands. At Pantex Plant these areas include the Plant perimeter and surrounding agricultural lands.

8 Conclusions

Environmental restoration and cleanup of damaged ecosystems must be driven by risk-based decisions rather than perception. Various cleanup options are expensive, and some may cause greater environmental damage than if cleanup were never attempted. At Hanford, Pantex Plant, and other DOE sites, cleanup will be a significant technical and management challenge, greater than that experienced 50 years ago when the sites were selected and built. In contrast to that earlier time, cleanup will be conducted under the scrutiny of a doubting public, a critical press, environmental activists, and governmental and other oversight groups.

Environmental monitoring for radiological emissions at the Hanford Site and Pantex Plant has been conducted for several decades and includes air, surface and ground waters, foodstuffs, fish, wildlife, soils, and vegetation. At both sites, measured and calculated radiation doses to the public have consistently been below applicable regulatory limits. At the Hanford Site, monitoring of fish and wildlife populations and characterization and protection of archeological and cultural resources are also significant components of the programme. The Hanford Site now serves as a refuge for various plant and animal species. Similar studies now being planned will determine if this situation also occurs at the Pantex Plant. Stable or growing populations of key biotic species resulting from Hanford's 50 years of protection from public intrusion are consistent with the low radiological doses. Even lower doses are reported at Pantex Plant. Given these data, reasonable people may ask why clean these sites up at all ?

While it may appear that public perception and politics are driving the scientific process, certain cleanup and remedial actions are essential. The degree of cleanup required has yet to be determined; some cleanup options may cause more harm than no cleanup, and some cleanup actions may be unnecessary, particularly where public safety and health risks are low. If the future calls for opening the Hanford Site to agricultural or other public use, its value as an ecological study area, environmental research park, Nature Conservancy or archeological and cultural resource study area will be lost. Eventually the taxpayers should have the opportunity, through election or influence of their legislative representatives, to decide if the cleanup efforts is worth the expenditure of billions of dollars. Their decisions must not be driven by distortions, misrepresentations, and unfounded fears. At the same time, some guarantee of protection for future generations must be provided. Meeting these challenges is central to developing coordinated and consistent approaches to determining the future of these ecosystems.

Acknowledgements

I thank Laurel Grove, Bob Pankratz, and Carl Phagan, who critically reviewed the manuscript. This work is supported by DOE under Contract DE-AC06-76RLO 1830 with Battelle Memorial Institute and Contract DE-AC04-91AL-65030 with Mason and Hanger - Silas Mason Co, Inc.

References

1. R. H. Gray, *Environ. Mon. Assess.* **26**, 263 (1993).
2. R. H. Gray, and D. A. McGrath, *Fed. Fac. Environ. J.* **6**, 79 (1995).
3. J. C. Chatters, PNL-6942, Pacific Northwest Laboratory, Richland, WA. (1989).
4. F. B. Largent, Mason and Hanger, Amarillo, TX. (1995).
5. D. C. Stapp, J. K. Woodruff, and T. E. Marceau. *Federal Archaeology* **8**, 14 (1995).
6. DOE. DOE/S-0070, U.S. Department of Energy, Washington, DC. National Technical Information Service, Springfield, VA. (1989).
7. DOE. DOE/S-0089, U.S. Department of Energy, Washington, DC. National Technical Information Service, Springfield, VA. (1991 a).
8. DOE. U.S. Department of Energy, Richland, WA. (1991 b).
9. R. H. Gray, R. E. Jaquish, P. J. Mitchell, and W. H. Rickard, *Environ. Manage.* **13**, 563 (1989).
10. R. E. Jaquish, and R. W. Bryce (eds.) National Technical Information Service, Springfield, VA. (1990).
11. R. K. Woodruff, R. W. Hanf, M. G. Hefty, and R. E. Lundgren, (eds.) PNL-7930, Pacific Northwest Laboratory, Richland, WA National Technical Information Service, Springfield, VA. (1991).
12. R. K. Woodruff, R. W. Hanf, M. G. Hefty, and R. E. Lundgren, (eds.) PNL-8148, Pacific Northwest Laboratory, Richland, WA National Technical Information Service, Springfield, VA. (1992).
13. R. L. Dirkes, R. W. Hanf, R. K. Woodruff, and R. E. Lundgren, (eds.) National Technical Information Service, Springfield, VA. (1994).

14. DOE. DOE/AL/65030-9413, National Technical Information Service, Springfield, VA. (1994).
15. DOE. DOE/AL/65030-9506, National Technical Information Service, Springfield, VA. (1995).
16. L. E. Bisping, National Technical Information Service, Springfield, VA (1995).
17. W. D. McCornmack, and J. M. V. Carlilel, PNL-5289, Pacific Northwest Laboratory, Richland, WA National Technical Information Service, Springfield, VA. (1984).
18. K. R. Price, National Technical Information Service, Springfield, VA. (1986).
19. EPA. EPA-570/976-003, U.S. Environmental Protection Agency, Washington, D.C. (1976).
20. R. L. Dirkes, National Technical Information Service, Springfield, VA. (1988).
21. C. E. Cushing, D. G. Watson, A. J. Scott, and J. M. Gurtisen. *Health Phys.* **41**, 59 (1981).
22. J. H. Honea, and T. A. Gabocey, National Technical Information Service, Springfield, VA. (1991).
23. R. L. Dirkes, G. W. Patton, and B. L. Tiller, National Technical Information Service, Springfield, VA. (1993).
24. E. H. McGavok, W. P. Wiggins, R. L. Blazs, P. R. Boucher, L. L. Reed, and M. L. Smith, U. S. Geological Survey, Tacoma, WA. (1988).
25. C. D. Becker, and R. H. Gray, *Environ. Mon. Assess.* **22**, 137 (1992).
26. WSDOE. Washington State Department of Ecology, Olympia, WA. (1977).
27. R. H. Gray, and W. H. Rickard. *Environ. Conser.* **16**, 250 and 215 (1989).
28. R. E. Fitzner, and R. H. Gray, *Environ. Mon. Assess.* **18**, 173 (1991).
29. W. H. Rickard, and R. H. Gray, Natural Areas J. **15**, 68 (1995).
30. R. H. Gray, and D. D. Dauble, Northw. Sci. **51**, 208 (1977).
31. D. G. Watson, BNWL-1750, Pacific Northwest Laboratory, Richland, W.A. National Technical Information Service, Springfield, VA. (1973).
32. G. A. Swan, E. M. Dawley, R. D. Ledgerwood, W. T. Norman, W. F. Cobb, and D. T. Hartman. National Marine Fisheries Service, National Oceanic and Atmospheric Administration, Seattle, WA. (1988).
33. W. C. Hanson, and L. L. Eberhardt, *Wildlife Soc. Monograph No.* **28**, (1971).
34. R. E. Fitzner, L. E. Eberhardt, W. H. Rickard, and R. H. Gray, *Northw. Sci.* **68**, 37 (1994).

35. W. H. Rickard, J. R. Hedlund, and R. E. Fitzner, *Science* **196**, 1009 (1977).
36. W. D. Steigers, and J. R. Flinders, *J. Wildlife Manage.* **44**, 381 (1980).
37. C. E. Cushing, R. R. Mazaika, and R. C. Phillips National Technical Information Service, Springfield, VA. (1993).
38. M. C. Johnston, and J. K. Williams, U. S. Department of Energy, Amarillo, TX. (1993).
39. M. K. Rylamder, U. S. Department of Energy, Amarillo, TX. (1994).
40. K. D. Seyffert, Contract No. FFP016902, U. S. Department of Energy, Amaraillo, TX. (1994).
41. M. C. Johnston, U. S. Department of Energy, Amarillo, TX. (1995).
42. N. M. Stricker, and R. M. Poet, Legacy Research Association, Inc., Corvallis, OR. Battelle Pantex, Amarillo, TX. (1994).
43. C. A. Geffen, B. A. Garret, C. E. Cowan, M. R. Siegel, and J. R. Keller, PNL-6972, Pacific Northwest Laboratory, Richland, W.A. National Technical Information Service, Springfield, VA. (1989).
44. R. H. Gray, and C. D. Becker, *Environ. Manage.* **17**, 461 (1993).
45. DOE. U.S. Department of Energy, Washington, DC. National Technical Information Service, Springfield, VA. (1990).
46. R. H. Gray, in *Effective and Safe Waste Management : Interfacing Sciences and Engineering with Monitoring and Risk Analysis*, eds. R.L. Jolley and RGM Wang Lewis Publishers, Boca Raton, FL. (1992).
47. C. D. Becker, National Technical Information Service, Springfield, VA. (1993).
48. C. D. Becker, *Studies in Environmental Science 39*. Elsevier Science Publishing, New York, NY (1990).
49. W. T. Farris, B. A. Napier, T. A. Ikenberry, J. C. Simpson, and D. B. Shipler, PNWD-2229 HEDR, Battelle Pacific Northwest Laboratories, Richland, W.A. National Technical Information Service, Springfield, VA. (1994a).
50. W. T. Farris, B. A. Napier, J. C. Simpson, S. F. Snyder, and D. B. Shipler, 1944-1992, PNWD-2227 HEDR. Battelle Pacific Northwest Laboratories, Richland, W.A. National Technical Information Service, Springfield, VA. (1994b).
51. D. A. Neitzel, T. L. Page, R. H. Gray, and D. D. Dauble, *Environ. Impact Assess.* **3**, 43 (1982).
52. R. H. Gray, *Fisheries* **16**, 2 (1991).

53. NPS. Nationla Park Service, U. S. Department of the Interior, Seattle, W. A. (1992).

ECOSYSTEM RESTORATION : A NEW PERSPECTIVE FOR SUSTAINABLE USE OF THE PLANET

JOHN R. HECKMAN AND JOHN CAIRNS, JR.

Department of Biology, Virginia Polytechnic Institute and State University, Blacksburg, VA 24061-0415, USA

1 Introduction

Fragmentation, urbanization, and other human-caused disturbances of ecosystems are negatively impacting the biological support system upon which all human society ultimately depends[1]. It has been estimated that over 40% of Earth's productivity has been modified and harnessed for human ends[2]. In many cases, the most productive or key ecosystems are the ones that have been the most impacted. In North America, wetland acreage of all types has decreased more than 50% over the past two centuries[3]. If sustainable use of the planet is the goal, these historical precedents and the continuing trends towards urbanization and habitat destruction make a convincing case for a highly integrated programme of ecosystem restoration and renewal[4].

Current restoration efforts usually do not focus on large-scale processes and, therefore, do not ensure reintegration into the greater landscape[5]. The largest proportion of restoration projects are attempts to reconstruct a specific "historical ecosystem", and success is judged by measures of gross community structure which are crude measurements for how the restored patch interacts with its surrounding ecosystem[6]. This view is at odds with current theories of ecosystem ecology and echoes a Clementsian view of climax succession. Ecosystems are not homeostatic, with a specific set point for which all succession aims[7]. Rather, ecosystems are homeorheoic; they are dynamic, feedback systems that fluctuate within a range of states[8]. Restoration of ecosystems can be more effective if the larger scale processes involved in feedback loops are taken into account on smaller scales, even if these processes seem to be unimportant at that scale[9]. One way this can be effected is by defining specific ecosystem functions that are expected to be performed by ecosystems of different types. By studying the feedback communication between ecosystem components and understanding the effect that disturbance and restoration have on these interactions, it may be possible to work from within ecosystem processes to guide recovery. This performance-based judgement process can then be used in conjunction with economic and sociological considerations for making restoration decisions.

2 The Problem-solving Perspective

The need to restore ecological systems arose from the increase in environmental damage that occurred during the industrial revolution. Current restoration and reclamation practices represent the culmination of research and practical experience in dealing with these impacts[10]. This experience has produced a well-rounded suite of techniques that is useful and necessary for dealing with the problems associated with human activity. However, along with this highly capable stable of site-specific techniques, the restoration lineage has created a perspective that views restoration efforts on a small scale. Because a problem may have specific confines and dimensions, solutions are addressed and judged by the same dimensions. This usually means judgement based upon what can be observed at the area that incurred the impact. For example, Lake Michigan was considered polluted in the late 19th century due to direct dumping of sewage from the city of Chicago. In 1900, the construction of the Chicago Sanitary and Ship Canal diverted sewage from the lake into the Mississippi River, and the lake was considered "restored"[3]. Similarly, in the 1960s and 1970s, eutrophication due to nutrient enrichment from sewage treatment plants contributed to severe crashes in populations of commercially important fish[11]. This problem was addressed by governmental regulations of phosphorous loading and, again, the lake was considered "restored".

This problem-solving perspective of restoration was extremely successful for dealing with the severe impacts associated with industrial activity before environmental regulations became widely applied. However, as community and ecosystem ecologists have learned more about the effects of pollution, habitat destruction, and other environmental damages, a single problem/single solution perspective is obviously not sufficient. Evidence for this inadequacy lies in problems with designing restoration plans to deal with the chronic effects of toxicants and long-term, large-scale effects (Table 1).

Table 1: Environmental problems that resist a problem-solving approach for restoration

Problem	Ramifications	Reference
Developmental toxins	Compounds released into the environment can, in very small amounts, interfere with hormonal activities, thereby interrupting normal animal development and seriously impacting reproductive fitness	16,17
Habitat fragmentation	Destruction of natural communication corridors severely impedes animal population dynamics, plant pollination and migratory routes necessary for climate change adaptation. This is confounded by the dependence of humans on the same fragmentation	18
Loss of biodiversity	Reduction of genetic potential impedes recolonization of disturbed areas, impacts as yet unknown ecological processes, and permanently removes the opportunity for human or other uses of many organisms	19
Increased atmospheric CO_2	The unprecedented rapidity of the recent increase in atmospheric CO_2 may have unpredictable effects on primary productivity, solar warming, and climate changes. The long turnover time of the atmospheric CO_2 pool makes the effects of any restoration project long term at best	20

Such effects, coupled with ecological flux and dynamics, hinder the prediction of ecosystem performance after impact. Restoration ecologists are starting to deal with the problem of judging ecological performance of restored or recovering systems[12,13]. The question still remains, however : How is this information incorporated into the realistic action of ecological restoration?

3 An Alternative View : Cybernetic Restoration

The key to developing an organized approach towards restoring ecological structure and function may be to step back from the problem-solving perspective, which has dominated restoration theory, and view disturbed areas as components of an impaired system. Ecosystem ecologists have shown that natural biotic systems can be seen as cybernetic or positive feedback systems[14]. As shown in Figure 1a a cybernetic system is partially or entirely dependent upon its output to control its input. The restoration of this transfer of information between subsystems is crucial for restoring a self-maintaining ecosystem. Many restoration attempts, however, pay very little attention to these informational flows and instead, effectively focus on the independent recovery of the subsystems. When viewed from the problem-solving perspective, i.e., considering the problem to be the destruction or alteration of one of the subsystems themselves, this approach works very well. In the Lake Michigan example, the fish population subsystem was restored to an economically satisfactory level.

With the most pressing environmental problems today, such as those shown in Table 1, it is the informational flows between subsystems that are most important and, therefore, must be included in the goals of any restoration efforts. In such cases, efforts should focus on interpreting the informational flows and altering the subsystem performance to meet an end goal of ecosystem self maintenance (Figure 1b).

A.

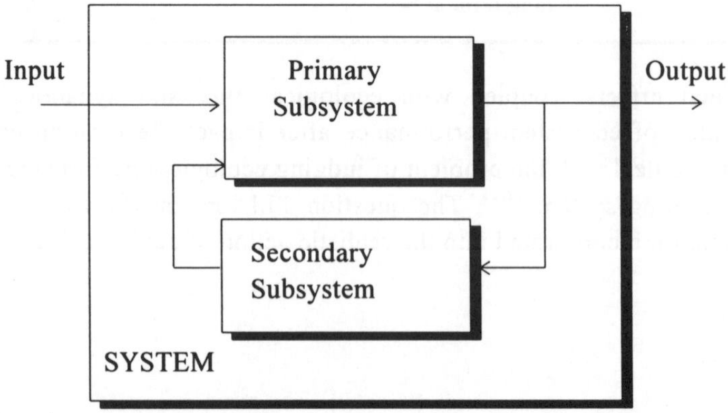

Figure 1A : Cybernetic model of an ecosystem showing feedback.

B.

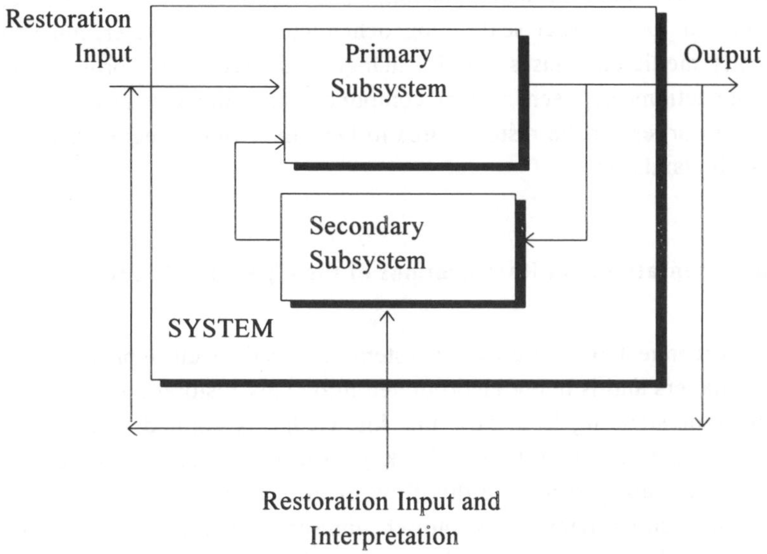

Figure 1B : Interpretation and control of feedback information during ecosystem restoration. (Modified from[14]).

The current problem of developmental toxin pollution provides a simplified example. In this situation, the human society subsystem is effectively sending information to the multitude of mammal, fish, and bird systems, which instructs them to alter their developmental patterns destructively. The subsystem performance is important to this problem, but the information shared between the systems is at the basis of the problem.

The importance of focusing on restoring system interactions can be better exemplified on a smaller scale. Consider a landscape severely impacted by strip-type mining. All restoration logic will place priority on the recovery of the plant community. Beyond this, however, monitoring the informational flows is the best method for understanding the manner in which the restoration should proceed and the way in which the land can best be reintegrated into the landscape. The plant community of the area is open to new input from widely distributed weed species that require bare ground for establishment. The recovering vegetation provides cues to returning wildlife based upon what food sources are available. The recolonization process further affects the external wildlife populations by providing more free range and thereby altering population pressures and predator/prey dynamics. Edge effects and other emergent properties of heterogeneous landscapes also add to the feedback information. On a larger scale, biogeochemical flows of water, nutrients, and biotically-mediated gases will change as recovery progresses. The interconnections between system components are the key aspects that must recover in order for the restored area to become a well-integrated, functioning part of the landscape.

4 Implementation and Ramifications of the Cybernetic Perspective

This interconnected nature of ecosystem is already well appreciated by most land managers and is being incorporated into many restoration concepts [13]. The gap in understanding is in how this knowledge is applied to the restoration process in a practical setting. The key aspect for accomplishing this is to actively intercept and decode this flow of information in an effort to modify the various subsystems on a shorter timescale than the entire system can modify itself based upon the feedback loops (Figure 1 b).

This detailed monitoring process puts restoration practitioners inside the web of ecosystem processes and gives them the effective role of teleological controller. While ecologists may be quick to note that ecosystems lack any sort of telos, or end goal, restoration situations absolutely require a telos that incorporates human needs and natural tendencies. Unless human society learns how to incorporate its actions within the natural processes themselves, restoration efforts will be sub-optimal, show slow response to emergent problems, and fail to return the desired result.

To implement this restoration approach, it is necessary to identify the subsystems of interest, to identify and understand how the components interact, and to develop methods to monitor and act upon the informational flows. In other words, ecological knowledge and naturalistic data gathering must be employed as an integral component of any restoration plan. A wealth of information already exists concerning ecological interactions, from the necessary characteristics of bird nesting habitat to the historical ranges of plant species and their associated animal populations. The challenge is to create restoration protocols that efficiently apply this information, as well as planned monitoring of new information throughout the restoration process. This is necessary so that the benefits of ecological knowledge can be fully utilized while limiting the ecological research to within the constraints of a restoration project budget.

The 21st century has been called the Age of the Environment, marked by a new appreciation for the services human society is getting by the ecological support system[21]. If this designation is to be true, the restoration process must be optimised and integrated into all economic activity. The multivariate and dynamic properties of nature make this integration a difficult task. Therefore, an ordered, scientific approach is required to demonstrate how human society's actions can best be incorporated into ecological processes. To be sure, the problem-solving approach towards restoration must be modified so that it takes into account the cybernetic nature of ecosystems. This step may be the first in creating a process of true ecological restoration.

References

1 J. Cairns Jr. Water Science and Technology Board, National Research Council, Washington, DC (1994).

2 P. M. Vitousek, P. R. Ehrlich, A. H. Ehrlich, P. A. Matson, *BioScience* **36**, 368 (1986).

3 National Research Council (NRC). *Restoration of Aquatic Ecosystems Science, Technology, and Public Policy.* National Academy Press, Washington, DC (1992).

4 A. R. E. Sinclair, D. S. Hik, O. J. Schmitz, G. G. E. Scudder, D. H. Turpin, N. C. Larter, *Ecological Applications* **5**, 579 (1995).

5 J. Cairns Jr., J. R. Heckman, *Annual Reviews of Energy and Environment* **21**,167 (1996).

6 S. T. A. Pickett, V. T. Parker , *Restoration Ecology* **2**, 75 (1994).

7 E. P. Odum, *Science* **164**, 262 (1969).

8 M. Westoby, B. Walker, I. Noy-Meir, *Journal of Range Management* **42**, 266 (1989).

9 Z. Naveh, *Restoration Ecology* **2**, 180 (1994).

10 J. A. Harris, P. Birch, J. Palmer, Addison Wesley Longman, Essex (1996).

11 W. J. Christie, *Journal of the Fisheries Research Board of Canada* **31**, 827 (1974).

12 J. Aronson, C. Floret, E. Le Floc'h, C. Ovalle, R. Pontanier, *Restoration Ecology* **1**, 168 (1993).

13 R. J. Hobbs, D. A. Norton, *Restoration Ecology* **4**, 93 (1996).

14 B. C. Patten, E. P. Odum, *American Naturalist* **118**, 886 (1981).

15 A. Gore, *Earth in Balance, Ecology and the Human Spirit.* Houghton Mifflin, Boston (1992).
16 T. Colburn, C. Clement in *Advances in Modern Environmental Toxicology,* M. A. Mehlman, (ed.), Princeton Scientific, Princeton, NJ (1992).
17 T. Colburn, D. Dumanoski, J. P. Myers, Penguin, New York (1996).
18 D. A. Saunders, R. J. Hobbs, P. R. Ehrlich, Surrey Beatty and Sons, Chipping Norton, New South Wales, Australia (1993).
19 E. O. Wilson, *Biodiversity.* National Academy Press, Washington, DC (1986).
20 G. M. Woodwell, F. T. Mackenzie, Oxford University Press, New York (1995).

FREE-FLOATING HYDROPHYTES FOR TREATMENT OF WASTEWATERS

Y. B. HO AND J. S. H. TSANG

Department of Botany, The University of Hong Kong, Hong Kong

1 Introduction

The use of aquatic plants in the treatment of wastes in south-east Asia dates back many centuries. In southern China traditionally human and agricultural wastes have been recycled through integrated polyculture dike-pond systems by which energy and nutrients are recycled in a very complex manner[1]. Scientific study of the use of aquatic plants for wastewater treatment began in Germany in the 1950s during which Seidel and co-workers made use of *Scirpus* and *Phragmites* to treat dairy wastewater[2]. In North America many studies were initiated in the 1970s on the use of aquatic systems in the treatment and recycling of various wastes[3].

Both natural wetlands[4] and man-made[5] aquatic systems have been utilized in wastewater treatment. Natural wetlands are very complex ecosystems and since they are not designed for such function their treatment capability is very difficult to predict. On the other hand man-made aquatic systems are designed and built for wastewater treatment and thus prediction on and control over their functional capability are more feasible[5].

There are many types of man-made aquatic treatment systems and a number of ways to classify them. One way is to classify them according to the life form of the dominating hydrophytes in the systems[6]. Accordingly four major types of systems are classified, namely, emergent hydrophyte treatment system, submersed hydrophyte treatment system, submersed hydrophyte treatment system, free-floating hydrophyte treatment system, and integrated hydrophyte treatment system. Emergent hydrophyte treatment system are widely used in temperate regions. Common emergent hydrophytes selected include *Phragmites australis*[7], *Scirpus sp.*[8], *Typha latifolia*[9] and many others. The growth of plants in water is much influenced by light penetration and hence submerged hydrophyte treatment systems operate best with effluents low in turbidity such as secondary treated wastewaters. Submerged plants that favour high nutrient media are normally most suitable for treating wastewaters. *Enteromorpha* sp. (macroalgae), *Sphagnum* sp. (moss), and various species of the genera *Myriophyllum, Elodea, Ceratophyllum* and *Potamogeton* (higher plants) are generally considered to be suitable[10,11]. Free-floating hydrophyte treatment systems are widely used for treating and purifying various wasterwaters. Some free-floating hydrophytes such as *Eichhornia crassipes* and *Pistia stratiotes* are highly productive and form the dominate plants in the treatment systems in the subtropics and tropics[12] Finally the integrated hydrophyte treatment systems are a combination of any two or more of the above-mentioned systems or, in addition, with any other conventional low-technology treatment systems[5]. Two recent examples of the use of such systems are those of Triet *et al.*[13] and Johengen and Larock[14].

This paper reviews on the free-floating hydrophyte treatment system which is the major hydrophyte treatment system in the tropics and subtropics. Earlier reviews on the use of wetland plants to treat wastewaters includes those by Athie and Cerri[15], Reddy and Smith[16] and Gray[17]. More recent reviews, particularly on the use of emergent and integrated treatment systems, are those of Moshiri[18], Kadlec and Brix[19] and Kadlec and Knight[20]. Trivedy and Gudekar[21] reviewed specifically on the use of water hyacinth *Eichhornia crassipes* for wastewater treatment.

2 Design of the Free-Floating Hydrophyte Treatment System

It is not the intention of this paper to review in depth the design of different floating hydrophyte treatment systems (FHTSs). Readers could find such details in Tchobanoglous[22] and Reed *et al.*[23]. The actual design and operation of FHTSs depend on a number of factors such as the characteristics of the wastewater to be treated and the quality of the effluent required, the hydrophytes used in the system, and the local environmental conditions such as the climate. Typically a FHTS consists of a single[24] or a series[12] of shallow ponds with a water depth normally less than one metre, and with the stocked hydrophytes growing on the water surface. The ponds are usually elongated with the wastewater to be treated entering one end and, after flowing through the whole length of the hydrophyte mat, leaving on the other end of the pond. In some cases the wastewater is recycled to obtain a better quality effluent[25]. Most FHTSs operate with a hydraulic detention time of 5 days or longer.

Although a few cases have been reported[26] in using FHTSs to treat primary effluents most are designed for secondary effluents which are characteristically having lower BOD_5 and TSS (total suspended solids) levels. For treatment of primary effluents the system would operate anaerobically due to the high organic content of the effluent. Table 1 gives some suggested design criteria for FHTSs.

Table 1 Suggested design criteria for free-floating hydrophyte treatment systems

	Primary Treatment System	**Secondary Treatment System**
Design criteria		
Areas of each pond	0.4 ha	0.4 ha
Maximum depth	<1.5 m	<1.0 m
Length/width ratio	>3:1	>3:1
Hydraulic loading rate	500 m^3 ha^{-1} day^{-1}	800 m^3 ha^{-1} day^{-1}
Hydraulic residence time	>50 days	>5 days
Organic loading rate (BOD)	<200 kg ha^{-1} day^{-1}	<100 kg ha^{-1} day^{-1}
Nitrogen loading rate	<20 kg ha^{-1} day^{-1}	<15 kg ha^{-1} day^{-1}
Influent flow diffuser	Desirable	Desirable
Mosquito control	Essential	Essential
Plant harvesting	Seasonal to annual	Weekly to monthly
Effluent requirements		
BOD_5	<30 mg l^{-1}	<10 mg l^{-1}
Total suspended solids	<50 mg l^{-1}	<10 mg l^{-1}
Total nitrogen	<20 mg l^{-1}	<10 mg l^{-1}
Total phosphorus	<7 mg l^{-1}	<5 mg l^{-1}

Modified from O'Brien[27] and Tchobanoglous[22].

3 Operation Mechanism of FHTS

As FHTS is mainly designed for treating wastewater at the secondary and/or tertiary level, in most cases its function is for the reduction of suspended solids and BOD in, and removal of nutrients, in particular, nitrogen and phosphorus from wastewaters. One can arbitrarily divide the system into three parts, namely, the free-floating hydrophyte layer, the clear water zone, and the sediment layer at the bottom. Each part is important and plays its respective role in wastewater treatment.

In the free-floating hydrophyte layer the roots and other submerged organs absorb from the wastewater effluent nitrogen, phosphorus and other contaminants such as metals and organic compounds. So hydrophytes having extensive root system, high growth rate, fast uptake and luxurious storage of nutrients, and tolerance to high levels of contaminates in wastewater are selected to stock the FHTS. Roots of hydrophytes also oxygenate the bathing medium with air from the shoot[28]. Thus the root system creates a microenvironment in which various microorganisms, either adhering on the root surface or in the water, oxidize organic carbon hence obtaining energy for growth, and deaminate organic nitrogen to ammonia which is then nitrified to nitrate. Conversely there are also locations where oxygen depletion occur and these are sites where nitrogen may be lost from the system through denitrification[29]. The root system also helps to adsorb and sediment suspended solids from the wastewater. Regular harvest of the hydrophytes would remove the absorbed and stored nutrients and other substances from the treatment system.

The clear water zone accomodates the bulk of the wastewater in the pond. It supplies and exchange with wastewater in the root system of the hydrophytes. Such as flux of wastewater, from the clear water zone to the root layer where it is treated, plays a very important part in the treatment process[30]. Further, depending on the redox potential of the wastewater in the clear water zone, the microbes may oxidize or reduce organic carbon to gaseous forms before leaving the system. Similarly nitrogen in the wastewater may undergo metabolic processes such as deamination, nitrification, denitrification, ammonification and then either absorbed and metabolized by the microbes and hydrophytes or volatized and released into the atmosphere. Phosphorus may be absorbed and then metabolized, or get precipitated with cations.

Much of the suspended solids in the wastewater sediment out of the clear water zone and settle onto the bottom of the pond forming a sediment layer, especially towards the influent side. As diffusion of dissolved gases through water is very slow, rapid depletion of oxygen in the sediment occurs resulting in a sharp reduction in the redox potential 1-2 cm down the sediment profile.

This results in a thin oxidizing layer of sediment on top of a reducing layer below which obligate anaerobic bacteria thrives. Between these two layers a nitrification-denitrification reaction chain develops in that ammonium ion from the lower layer diffuses to the upper oxidizing layer where it becomes nitrified to nitrate, which in turn diffuses to the lower reducing layer and is denitrified to nitrogen and escapes from the system[31]. Phosphorus in the wastewater may be precipitated with iron and aluminium in the sediment.

4 Treatment of Municipal Wastewater

In the treatment of domestic wastewater, various components of the effluent including BOD_5, total suspended solids, nutrients, especially nitrogen and phosphorus, pathogenic microorganisms are reduced. The ways by which FHTS reduce these components in wastewater are sumarized as follows :

4.1 BOD_5

This basically reflects the organic carbon content in the effluent and its removal is mainly effected by mocroorganisms in the root and sediment zones of the FHTS. Organic carbon forms the energy source and building blocks of both the aerobic and anaerobic microorganisms in the treatment system. Reddy and Debusk[32] showed that a source of electron acceptors is crucial to the breakdown and hence removal of organic carbon from the effluent. For aerobic bacteria oxygen is the electron acceptor whereas for anaerobic bacteria, compounds such as nitrate and sulphate act as the acceptor. In an efficient FHTS the percentage and rate of removal of BOD_5 reaches 80-90% and 150-200 kg BOD_5 ha^{-1} day^{-1}, respectively.

4.2 Total Suspended solids

Removal of total suspended solids is mainly by gravity sedimentation onto the bottom of the pond although some particles adhered onto the submerged parts of the hydrophytes and subsequently removed during harvest. The plants also played an indirect role in sediment removed since their presence would reduce wind-blown turbulence and speed up the sedimentation process[23]. Efficiency in the reduction of suspended solids in FHTS may be higher than 80%.

4.3 Nitrogen

The ability of the hydrophytes to remove nitrogen from wastewater depends on the growth rate of the plant and its nitrogen uptake rate. There are considerable variations in the percentage removal of nitrogen by hydrophytes among different FHTSs studied and it might reach nearly 90%[33]. However in many cases the percentage removal was much lower, at 10% or below[34]. Most of the nitrogen is removed by microorganisms through nitrification, denitrification and ammonia volatization processes in the sediment and root zone of the pond.

4.4 Phosphorus

Similar to nitrogen, phosphorus in the wastewater might be taken up by the hydrophytes, incorporated in the microbes, or precipitated together with metals such as calcium in the sediment layer. However phosphorus removal efficiency on the whole is less than that of nitrogen since the element, unlike nitrogen, cannot be readily be converted to gaseous forms and escape from the treatment system[35].

4.5 Microorganisms

Various studies indicated that FHTS generally can remove total and faecal coliform, and other bacteria, including pathogenic ones, as well as viruses from wastewater[36,37]. Often the removal efficiency is very high and may reach 98% or more. Sometimes even parasitic organisms such as helminth eggs are also reduced. There are a number of mechanisms responsible for the removal of bacteria. They may adsorb onto the root and other submerged surfaces of the hydrophytes. Secretion of antibacterial compounds from the hydrophyte root may suppress the number of bacteria. Further the microbes may adhere onto the suspended solids which subsequently settle as sediments on the bottom of the pond. Natural die-off and the reduction of organic carbon and nutrients may also account for the reduction in the number of bacteria in the treatment system.

5 Treatment of Agricultural and Industrial Wastewaters

Apart from the treatment of municipal wastewater, free-floating hydrophytes have also been utilized for treating agricultural and industrial wastewaters. Examples of using FHTSs to treat various agricultural and industrial effluents are presented in Table 2. Similar to domestic wastewater, effluents generated from agroindustries normally have high COD and BOD loads and rich in nutrients.

The composition of effluents from industries can be very diverse and depends on the nature of the industry. For example, wastewater from oil refinery has high oil content, drainages from coal operation has high acidity, and effluents from film manufacturing, chemical, and tannery industries contain high level of metals. These chemicals when present in high concentrations may be toxic to hydrophytes. Thus often such effluents are preliminary or secondary treated first before final treatment with FHTOS. Or. alternatively, the industrial effluents are mixed and hence diluted with domestic wastewater before treatment[38].

For both agricultural and industrial wastewaters, COD, BOD, SS, and nutrients are reduced or removed in FHTSs in ways similar to those for domestic wastewaters. Metals in industrial wastewater may be absorbed by and accumulated in hydrophytes.

Table 2 Examples of using FHTSs for treating agricultural and industrial wastewaters.

Effluent type	Major Pollutant Removed	Hydrophyte	References
Agricultural wastes			
Pig waste	COD, BOD, SS, nutrients, microbes NO_3, COD, NH_4	*E. crassipes* *S. molesta*	34,39,40,50
Dairy waste	NH_4, No_3, PO_4	*L.minor*, *S. polyhiza*	41
Pisciculture	COD, BOD, SS, nutrients	*E. crassipes*	42
Sugar refinery	BOD, SS, BOD, nutrients	*E. crassipes* *Lemna* sp.	24, 43
Cheese manufacturing	BOD	*Lemna* sp.	44
Distillery waste	BOD	*E. crassipes*, *S. molesta*	45
Pectin	BOD, SS, NH_4, P	*E. crassipes*	34
Palm oil mill	SS, BOD, NH_4	*E. crassipes*	24
Rubber manufacturing	SS, BOD, NH_4	*E. crassipes*	46
Industrial wastes			
Coal mining	Total acidity	*E. crassipes*, *S. quiquefarium*	10
Metal industry	Metals	*A. filiculoides*	47
Chemical industry	Cu	*S. natans*	48
Match manufacturing	SS, COD, BOD, P	*Lemna* sp., *Azolla* sp.	49
Oil refinery	BOD, petroleum	*E. crassipes*	13, 51
Textile industry	COD, BOD, SS	*E. crassipes*	52
Film manufacturing	Ag, COD, BOD	*E. crassipes*	53
Tannery	COD, BOD, Cr, SS Cr	*E. crassipes* *S. polyrhiza*	54 55 56

Table 3 Metal accumulation by hydrophytes

Metal	Hydrophyte	References
Ag	*Eichhorinia crassipes*	53,57
Cd	*Eichhorinia crassipes* *Spirodela polyrhiza*	9, 58, 11
Cr	*Salvinia molesta* *Spirodela polyrhiza*	11, 59
Cu	*Lemna minor* *Salvinia molesta*	48, 60
Fe	*Eichhorinia crassipes* *Spirodela polyrhiza*	11,61
Mn	*Spirodela polyrhiza*	11
Ni	*Eichhorinia crassipes* *Salvinia molesta,* *Spirodela polyrhiza*	59, 62
Pb	*Azolla pinnata* *Eichhorinia crassipes* *Lemna minor*	48, 63, 64
Se	*Azolla caroliniana,* *Lemna minor,* *Salvinia rotundifolia*	65
Zn	*Azolla pinata* *Eichhorinia crassipes*	48, 61

Various studies showed that many free-floating hydrophytes have considerable capacity to take up metals from solution examples of which are summarized in Table 3. The accumulated metals normally are located in the root while the shoot has lower levels of the metals[61]. In the uptake of metal ions by the roots of hydrophytes, a biphasic pattern of uptake is apparent for many of the metals investigated, including Ag[57], Cd[58], Pb[63] and Zn[9]. Initially there is a rapid phase of uptake usually lasting for a few hours and is followed by a slower, but more prolonged uptake phase. In the first phase absorption resulted from the binding of metal ions to various negatively charged groups, mainly $R.COO^-$, located in the apparent free space (both the water free space and Donnan free space) of the cell wall. Since the binding mechanism is a physical process, such metal uptake is possible in both living and nonliving[66] root materials. The slow, prolonged phase of uptake corresponds to metal absorption into the cell[58]. Within the cell some of the metals probably bind to phytochelatins since there is a close relationship between the amount of sulphydryl groups present (as basic chemical components of phytochelatins) and the metal concentrations in the root tissue of water hyacinth[67].

There are numerous report on the role of the phytochelatins or metallothioneins for sequestration of excess metal in plant[68,69]. These metal-binding proteins are invariably expressed as plant constituents upon exposure of the plant to heavy metals[70]. The ability of the plant to uptake metal from industrial wastewater can therefore be enhanced by increasing the amount of intrinsic phytochelatins by genetic engineering. Additional copies of the phytochelatin genes can be inserted into the genome of the plant for collateral expression or mutations can be introduced into the regulatory region of the corresponding promoters for better production.

Apart from metals, various organic compounds, including refractory ones, are commonly present in industrial wastewaters. Such organic substances include hydrocarbons, pesticides, polychlorinated biphenyls (PCBs) and many others. However relatively few studies have been made on their toxicity to, and uptake, translocation, accumulation and breakdown by aquatic plants[71]. Pioneer study by Wolverton[72] showed that water hyacinth very effectively removed various halogenated hydrocarbons and other trace organics from wastewater. The plant is also efficient in absorbing petroleum from oil refinery wastewater[50,51]. Other examples of the uptake of refractory organic compounds by floating hydrophytes include organochlorine insecticides by *Spirodela polyrhiza* and *Lemna minor*[73], chlorinated and other phenolic compounds by *Eichhornia crassipes*[74] and *Lemna minor*[75]. Most of these organic compounds are persistent, lipophilic and relatively insoluble in water. Various factors that affect the absorption of these organic compounds by the hydrophytes are their concentration, solubility and partitioning character in water, the morphometry and the lipid content of the hydrophyte[71]. After taken into the plant, these organic compounds are often transformed to different products and become detoxified. For example water hyacinth takes up pentachlorophenol and converts it to ortho- and para-substituted chlorohydroxyphenols, -anisoles, and -veratroles as well as some partially dechlorinated products[74]. Several stress-induced enzyme systems including glutathione S-transferase, superoxide dismutase and ascorbate peroxidase increase in activities and are linked to the biotransformation and detoxification of the absorbed orgnic compounds[74]. A similar increase in enzyme activities resulted after *Lemna minor* takes in hexachlorobenzene[75].

Other than the hydrophyte, microorganisms in association with the floating aquatic plant species also play a role in the treatment of wastewater. The microbes are mainly involved in the degradation of various organic compounds, including many persistent haloorganics, in the surrounding environment. Since the microbes are mainly contained in the sediments of the wastewater solid at the pond bottom and the majority of the water milieu in the FHTS is rather anaerobic, the degradation of various organic and halogenated compounds such as aromatic hydrocarbon, chlorinated homocyclic and heterocyclic aromatics, chlorinated aliphatic compounds and polychlorinated biphenyls etc., can be mediated via anaerobic mechanism[76]. The bacteria involved include *Desulfomonile, Acetobacterium, Clostridium* and *Methanobacterium*. Aerobic processes are mainly confined to the root zone where decomposition of organics and haloorganics can be achieved via oxygenases[77]. The microbes involved contain *Pseudomonas, Alcaligenes, Moraxella, Acinetobacter* and *Arthrobacter*. Since the root zone provides a niche for the microbes to propagate they normally form a film of growth along the root structure. Natural microbes isolates from this zone may have a rather limited degradative power but the use of genetic engineering should be able to modify this restriction, hence increasing the efficiency of FHTSs.

6 Operational Problems of the Treatment Systems

Experience gained in the operation of FHTSs indicated that certain problems commonly arise and are summarized below:

6.1 Oxygen Depletion

Low level of dissolved oxygen in the treatment ponds may arise as a result of too high an organic and/or hydraulic loading to the system. Various environmental factors such as high temperature and absence of sunlight (at night), and overcrowding of the hydrophytes which leads to a high percentage of senescent plant materials and accumulation of detritus in the treatment pond, also account for the oxygen depletion[17]. More effective pretreatment of the wastewater, reduction in the organic and hydraulic loading to the system, thinning of the hydrophytes, and aeration of the pond are effective remedial measures to maintain the dissolved oxygen level. Aeration can readily be effected by passing compressed air through submerged perforated pipes that extend through the front end of the pond. The absence of pipes at the rear end allows sedimentation and settlement of suspended solids[17].

6.2 Odours

The sulphur-containing hydrogen sulphide and mercaptans, and skatoles are compounds that give rise to odour problem in FHTSs[22]. These compounds are generated and persist in anaerobic condition. Mercaptans and skatoles are produced by intestinal microorganisms and are components of faeces. Sulphates in the pond sediment may be reduced to hydrogen sulphide by anaerobic microorganisms. Thus the problem of odour is closely related to and arise from low oxygen conditions in the pond. Removal of odours could be accomplished by maintaining aerobic conditions in the treatment systems by adopting similar procedures as for the avoidance of oxygen depletion.

6.3 Insect Infestation

Mosquito and weevil may easily breed in FHTSs. Proliferation of mosquito not only is a nuisance, but may also act as a host to parasites such as malaria. Various types of mosquito fish, including *Gambusia, Axtyarax, Lebistes,* and *Macropodus* species have been introduced to FHTSs to feed on the mosquito larvae. However such fish cannot survive in anaerobic condition which may develop in some treatment systems as discussed above. Alternatively application of insecticides will remove the larvae[22, 26]. Weevil may infest the plants resulting in their poor development, weaken their resistance to fungal infection, and speed up the process of senescence and detritus accumulation. All these will lower the operational efficiency of the treatment system. Weevil may be controlled by the introduction of its natural predators including tree frogs and anoles[35].

7 Common Free-Floating Hydrophytes for Wastewater Treatment

There are a number of free-floating macrophytes commonly used for the treatment of wastewater. The following is a brief description of the characteristics of these hydrophytes and their use in FHTSs.

7.1 Water hyacinth

The water hyacinth, *Eichhornia crassipes* (Mart). Solms, is a hydrophyte native to South America. It was introduced to the tropics and subtropics all over the world near the end of the last century. The plant has several characters which make it very suitable for wastewater treatment. It grows and spreads rapidly since it has a high growth rate and is able to reproduce vegetatively by the outgrowth of buds on the leaf axils and rootstocks giving ramets. The ramets eventually break off from the parent plant to become new individuals. Water hyacinth also has a very high capacity to take up nutrients, especially nitrogen and phosphorus, from water and store them in its tissue. Thus the high growth rate, coupled with luxurious uptake of nutrients, form a basis of nutrient removal from wastewater. Regular harvest, which is performed weekly to once every few weeks, helps to maintain a high growth rate of the hydrophyte and remove the accumulated nutrients from the treatment system. Also water hyacinth is able to take in air from the atmosphere and transport it internally to the root and aerates the rhizosphere[78]. This enhances the oxidation of organic compounds and hence reduces the BOD of the wastewater. The high density of water hyacinth plants provides shading and reduces the growth of algae which is a source of suspended solids and BOD (as organic carbon) in the treatment pond[5]. Further, as indicated previously, water hyacinth is efficient in removing metals and organic compounds from wastewater. All these characters make water hyacinth very suitable for the treatment of various types of wastewater. In addition the leaf of water hyacinth has two types of phenol cells, one type locates in the palisade layer, being elongated in shape and contains phenolic compounds enclosed in globules. The other type is isodiametric, with the phenolic compounds as an amorphous mass[79]. Martyn and Cody[79] detected various phenolic compounds including protocatechuic, caffeic, vanillic, *p*-coumaric and chlorogenic acids in the phenol cells. These compounds have anti-microbial, alleopathic and deterrent properties[80]. This indicates that water hyacinth has defence against herbivores and pathogens, and the water-soluble allelopathic compounds may be released into the water and suppress the growth of other plants[81]. It has also been demonstrated that the root of water hyacinth may syntheize and release antialgal compounds including N-phenyl-2-naphthylamine, linoleic acid, glycerol-1, 9-12 (ZZ)-octadecadienoic ester[82] and N-phenyl-1-naphthylamine[83] which are more powerful in their anti-algal effects than the common algicide copper sulphate. The abilities for self-defense and allelopathy are also desirable characters for plants used in wasterwater treatment[17].

Although very successful as a free-floating hydrophyte for FHTSs in tropical and subtropical regions, water hyacinth has its limitations since the optimum range of temperature for growth is 25-30^0 C and the growth rate is greatly reduced when the temperature drops below 10^0 C. The plant cannot survive in freezing temperature. Thus for FHTSs in temperate areas the plant only functions well outdoor in summer or needs to be placed in greenhouses to maintain high growth performance all year round. Alternatively, water hyacinth may be replaced by other hydrophytes, such as pennywort[84,85], watercress and duckweeds, that are tolerant to low temperature during the cool season. Pennywort also has high growth rate and capacity to take up nutrients from wastewater[86]. Another limitation of water hyacinth is that its growth is much affected by salinity in excess of 1.6%[87] and it cannot survive salinity higher than about 6-8%.

7.2 Water lettuce

Similar to water hyacinth, water lettuce (*Pistia stratiotes L.*) is a tropical plant and hence its growth is much affected by low, near-freezing temperature. It spreads vegetatively as its stolons give off secondary rosettes which, when detached, from new individuals. The plant has numerous fine roots which trap sediments efficiently and harbour various microorganisms that play an important role in wastewater treatment. Water lettuce is a fairly large free-floating aquatic plant and has a very high productivity and capacity of taking up nutrients from water[85,88]. In these respects water lettuce is only second to water hyacinth among the floating hydrophytes commonly used for wastewater treatment (refer to Table 4). In the treatment of wastewater, water lettuce attains higher growth rates at lower plant density[88] since the plant grows horizontally and at high densities self-shading occurs.

Table 4 Productivity and nutrient removal capacity of free-floating hydrophytes*

Hydrophyte	Standing Crop (kg DM m^{-2})	Productivity (kg DM m^{-2} yr^{-1})	Standing Stock ($g\ m^{-1}$)		Removal Rate ($g\ m^{-2}\ yr^{-1}$)	
			N	P	N	P
E.crassipes	0.5-3	0.2-11	30-90	3-18	7-585	7-234
P.stratiotes	0.5-1	0.2-8	9-25	2-5.7	135-511	30-110
Lemna sp	0.01-0.3	0.6-5.5	0.4-5	0.1-3.	35-120	12-80
Spirodela sp	0.1-0.2	0.01-4.4	5.7	1.4	58	15
Azolla sp	0.01-0.1	0.4-1.1	3.4	1.5	36	16
Salvinia sp	0.3	03.-4.5	1.5-9.0	0.4-2.	35-170	9-45

* Data sources are Reddy and DeBusk[32,89] and Oron[90], and Vymazal[91]

7.3 Duckweeds

These are cosmopolitan species and include the great duckweed (*Spirodela* sp), duckweed (*Lemna* sp), and the water meal (*Wolffia* sp). Duckweeds are rather small floating hydrophytes are rather small floating hydrophytes and hence have much smaller standing crop than water hyacinth and water lettuce.
However under high nutrients and optimal temperature conditions they show very high specific growth rates, with their biomass doubling every 3-5 days. This resulted in their moderate nutrient removal capacity. Duckweeds are highly modified plants with or without simple roots. Thus the rhizosphere, if present, is of limited extent and importance in harbouring microoganisms for the treatment of wastewater. This is very different from that of a water hyacinth system. However the fronds of duckweeds are efficient in removing nitrogen and phosphorus from water[89,92] and hence have been utilized to recover nutrients from wastewater[90]. Due to their vegetative mode and rapid rate of growth duckweeds easily form a dense cover on the water surface of the treatment ponds and would have two major effects. First light penetration into the water is much reduced and thus inhibits the growth of phytoplankton[93] and the photosynthetic production of oxygen. Second a dense cover of duckweeds retards the diffusion of atmospheric air, including oxygen, across the water

surface into the water column giving rise to anaerobic condition in the treatment pond. This would encourage the growth of anaerobic bacteria and denitrification activities.

Using duckweeds in FHTSs has several advantages including their ease of harvest, their high nutrition value as feed for animals[90,94], and efficiency in removing metals from wastewater, e.g. Cd[11], Cr[59,60,93], Fe, Ni, Pb and Zn. Since less affected by low temperature duckweeds have the additional advantage of being operative year-round in treatment systems outside tropical and subtropical regions. Relative to water hyacinth, duckweeds grow less well in full-strength wastewater with a high organic load and hence they work best in polishing secondary-treated wastewater for the removal of nutrients. Further due to their small size they are easily shifted by strong winds into piles of thick mats thus reducing the efficiency of wastewater treatment. To remedy this, poles, ropes, or other physical barriers could be installed in the pond surface to restrict wind-induced movements. Or, alternatively, mixed culture, FHTSs, using both duckweeds and some other hydrophytes such as water hyacinth or *Azolla*[49] could be used together for wastewater treatment.

7.4 Water Ferns

The two water ferns that have been used for the treatment of wastewater are water velvet (*Azolla* sp) and water fern (*Salvinia* sp). Same as the duckweeds the water ferns have undergone microphytization and hence are relatively small in size. They also have similar productivity and nutrient removal capacity to those of the duck weeds (Table 4). Thus the water ferns possess desirable characters comparable to the duckweeds for wastewater treatment. *Azolla* is well-known for its symbiotic relationship with *Anabaena azollae* which fix nitrogen from air. However the high concentration of ammonium in sewage effluent suppresses the ability of *Anabaena* to fix nitrogen[95]. This helps to maintain the ability of *Azolla* to remove nitrogen in wastewater treatment. Both *Azolla* and *Salvinia* take up and accumulate high levels of Se in their tissue[65]. Apparently the two water ferns can be utilized to remove Se from wastewater rich in the element such as leachate from coal ash lagoons. Apart from Se the two plants are efficient in the removal of heavy metals from wastewaters[47,48,59].

8 Utilization of Hydrophytes Harvested from FHTSs

As hydrophytes with high growth rate and nutrient removal capacity are selected for FHTSs, considerable biomass from FHTSs is generated and such plant materials should be properly managed through utilization and/or disposal. A major consideration for the management of such biomass is the overall high water content of about 95% for most free-floating hydrophytes. This makes the biomass very bulky and watery and relatively difficult to manage. Various ways of utilization of biomass from hydrophytes have been investigated and practised and are summarized below.

8.1 Methane production

This basically involves the biological reduction of organic carbon of the hydrophyte by various methanogenic bacteria to methane which is a gaseous fuel. The biogasification system at the Walt Disney World wastewater treatment plant serves as an example for such a process. In this system, chopped and ground water hyacinth tissue is mixed with sludge from the primary wastewater clarifies and the mixture then anaerobically digested in tanks to produce methane as the final product. At steady-state operation upto 0.49 m^3 methane per kg volatile solids is produced[97]. Similar systems are now in operation in the United States and other countries[98].

8.2 Animal Feeds

The use of hydrophytes to feed animals is a major way of recycling of biomass generated from FHTSs[94]. Floating hydrophytes generally are rich in vitamin contents, and have a range of crude proteins of around 10 to 30% on a dry weight basis[99] which is comparable to land plants. Traditionally in some Asian countries such as China and India, water hyacinth has been fed to livestock and poultry, including pigs, cattles and ducks. The high water (about 95%) and mineral salt contents are two undesirable factors that dampens the use of hydrophytes for feeds. The former increases the cost of harvesting, handling, transport and storage of the plants as raw materials, the latter reduces the palatability of the feeds as well as the mineral balance of the fed animals. To reduce bulk the harvested large hydrophyte such as the water hyacinth is normally chopped into smaller size and dewatered by draining and sun-drying before it is fed to the animals. Alternatively after dewatering it can be mixed with other feeds such as citrus pulp, sugarcane molasses, corn, rice and wheat bran[100] or wastes such as pig[101] or poultry excreta, and then fermented in silos. The silage produced becomes a suitable feed for cattles[17]. Ensilage also improves the nutritional value and the palatability of the feed. Among the free-floating hydrophytes duchweeds are especially suitable as feeds because they have high protein contents and can easily be harvested and handled[43].

8.3 Compost and Mulch

The biomass harvested from FHTSs contains considerable amount of mineral nutrients which can be recycled for growing crops and other plants. Sometimes the harvested plants are directly applied to soil[102], but often they are ashen or composted before application[99,101]. Plant materials are usually sun-dried and then burnt to obtain the ash. Ashing has the advantage of reduction in bulk and hence increase in the concentration of minerals. However certain mineral elements, especially N and S, would be partially lost through volatization. Fresh hydrophytes have too high a water content and require partial dewatering before composting. Often, to adjust for a more appropriate moisture level and C:N ratio, other materials such as sludge, dried livestock manure, soil etc. are included before composting. Apart from its fertilizer role compost often is more important as a source of organic matter when applied to improve its properties by enhancing aeration, water retention capacity and cation exchange capacity.

After drying aquatic plants can be applied as mulch to the soil surface or worked into it. Mulch tends to reduce evaporation, prevent soil erosion, and enhance shading, thus forming a protecting layer for the growth of plants. It may also be applied to suppress the germination and growth of weeds[103,99].

8.4 Other Applications

Free-floating hydrophytes, in particular the water hyacinth, have also been utilized as a source of fibre for the manufacturing of paper and boards. Normally the petiole and stem of the water hyacinth produce an acceptable pulp. However, fibre from the water hyacinth has a low freeness (does not drain water readily) and thus is not suitable for manufacturing paper using high speed machinery. Further paper made from the plant generally is of a low quality since it has high shrinkage, low in tear and burst thresholds and tensile strength, and a "dirty" appearance. Gopal[103] reviewed on the use of water hyacinth for making paper and boards. When blended with other fibres such as that of bamboo, water hyacinth can produce a good quality greaseproof paper which is high in freeness and physical strength, and has high tear and burst thresholds[104].

A few free-floating hydrophytes are suitable for human consumption such as *Wolffia arrhiza* which is regarded as a vegetable and taken whole[99]. Water spinach (*Ipomoea aquatica*) is a popular vegetable in many parts of south-east Asia and the aquatic form has been utilized in the treatment of wastewater in Thailand. It has been suggested that the amount of harvested plant from the treatment system can satisfy the needs of the local community. However in the use of hydrophytes from FHTSs as food it is important to ensure that they do not pose any health hazard after consumption.

Other potential applications of hydrophytes taken from FHTSs include their use as a substrate for the culture of mushrooms as medicinal herbs for treatment of various health disorders[99], and as materials for weaving into various articles such as mats, bags etc.[103].

9 Future Prospects

The free-floating hydrophyte treatment system is most suitable for small communities or industries, with land readily available, and located in regions with a hot climate. Improvement in the efficiency of FHTSs would ensure their wider acceptance and successful implementation in wastewater treatment. The operation efficiency for effluent treatment could be enhanced by the incorporation of a series of different systems, or by including different components within a system. An example of the former case is that of Triet *et al.*[13] who made use of a water hyacinth, a *Chlorella*, and a reed pond in series for post treatment of petroleum wastewater. The first pond was employed for oil film removal, the *Chlorella* pond for bioxidation of organic compounds and the last for stabilization. For the latter case Yang and co-workers developed highly efficient treatment systems with a combination of bio-fixed film and aquatic plant components[40]. The bio-fixed film consisted of a layer of microorganism coating on porous, volcanic rocks while *Eichhornia* sp. and *Salvinia molesta* were stocked in the aquatic plant component. Another strategy to raise the efficiency of FHTSs is by genetic engineering of the genomes of the hydrophytes and microorganisms so as to increase their capability to absorb, break down, and/or detoxify various compounds in the effluent. Although genetic engineering have made great advances in the agricultural and medical fields, very little similar work, if at all, has been reported in the area of wastewater treatment. We strongly believe that such a line of approach will bring forth substantial advancement in the biotreatment of wastewater in the near future.

Acknowledgements

This work is supported in part by a University of Hong Kong grant (grant no. 372/162/6392) given to YBH.

References

1. K. Ruddle and G. Zhong, *Integrated Agriculture-aquaculture in South China: The Dike-Pond System of the Zhujiang Delta.* Cambridge University Press, Cambridge (1988).
2. K. Seidel, in *Biological Control of Water Pollution.* ed J. Tourbier and R. W. Pierson, Jr., Pennsylvania University Press, Philadelphia (1976).
3. B. C. Wolverton, R. C. McDonald and J. Gordon, NASA Technical Memorandum TM-X72724 (1975).
4. D. S. Nichols. *J. Water Pollut. Control Fed.* **55,** 495 (1983).
5. H. Brix, in *Constructed Wetlands for Water Quality Improvement.* ed. G. A. Moshiri, CRC Press, Boca Raton. (1993).
6. H. Brix, and H. H. Schierup. *Ambio* **18,** 100 (1989).
7. P. F. Cooper, J. A. Hobson and C. Findlater, *Water Sci. Technol.* **22,** 57 (1990).
8. J. A. Moore, S. M. Skarda and R. Sherwood, *Water Sci. Technol.* **29**, 241 (1994).
9. G. Blake, J. Gagnaire-Michard, B. Kirassian and P. Morand. in *Aquatic Plants for Water Treatment and Resource Recovery,* ed. K.R. Reddy and W.H. Smith, Magnolia Publishing Inc., Orlando. (1987).

10. M. B. Falbo and T. E. Weaks, *Econ. Bot.* **44**, 40 (1990).
11. U. N. Rai, S. Sinha, R. D. Tripathi and P. Chandra, *Ecol. Engineering* **5**, 5 (1995).
12. S. A. Bramwell, and P.V.D. Prasad, *J. Environ. Management* **44**, 213 (1995).
13. L. M. Triet, N. T. Viet, T. V. Thinh, H. D. Cuong and J. C. L. Vanburen, *Water Sci. Technol.* **23**, 1503 (1991).
14. T. H. Johengen and P. A. Larock. *Ecol. Engineering* **2**, 347 (1993).
15. D. Athie, and C. C. Cerri. *Water Sci. Technol.* **19**(10), whole issue. Pergamon Press, New York. (1987).
16. K. R. Reddy and W. H. Smith, *Aquatic Plants for Water Treatment and Resource Recovery.* Magnolia Publishing Inc., Orlando (1987).
17. N. F. Gray, *Biology of Wastewater Treatment.* Oxford University Press, Oxford (1989).
18. G. A. Moshiri, *Constructed Wetlands for Water Quality Improvement.* Lewis Publishers, Boca Raton (1993).
19. R. H. Kadlex and H. Brix, *Water Sci. Technol.* **32**, Elsevier Science Ltd, Oxford (1995).
20. R. H. Kadlec and R. L. Knight. Treatment Wetlands. CRC Press, Inc., Boca Raton (1996).
21. R. K. Trivedy and V. R. Gudekar, in *Current Pollution Researches in India* ed. R. K. Trivedy and P. K. Goel, Environmental Publications, Karad, pp (1985).
22. G. Tchobanoglous in *Aquatic Plants for Water Treatment and Resource Recovery* ed. K. R. Reddy and W. H. Smith, Magnolia Publishing Inc., Orlando (1987).
23. S. C. Reed, R. W. Crites and E. J. Middlebrooks, *Natural Systems for Waste Management and Treatment.* 2nd edition. McGraw-Hill, Inc., New York (1995).
24. B. G. Yeoh, *Water Sci. Technol.* **28**, 207 (1993).
25. Y. B. Ho and W. K. Wong, *Resources, Conservation Recycling* **11**, 161 (1994).
26. E. J. Santos, E.H.B.C. Silva, J.M. Fiuza, T.R.O. Batista and P. P. Leal. *Water Sci. Technol.* **19**, 25 (1987).
27. W. J. O'Brien, *J. Environ. Engineering Div.* - ASCE **107**, 681 (1981).
28. A. Jedicke, B. Furch, U. Saint-Paul and U. B. Schluter, *Amazoniana.* **11**, 53 (1989).
29. K. R. Reddy *J. Environ. Qual.* **12**, 137 (1983)

30. R. Stowell and G. Tchobanoglous. Department of Civil Engineering, University of California, Davis (1983).
31. B. J. Good and W. H. Patrick Jr. in *Aquatic Plants for Water Treatment and Resource Recovery* ed. K. R. Reddy and W. H. Smith, Magnolia Publishing Inc., Orlando (1987).
32. K. R. Reddy, T. A. Debusk, Water Sci. Technol. **19**, 61 (1987).
33. V. R. Joglekar and V. G. Sonar in *Aquatic Plants for Water Treatment and Resource Recovery* ed. K. R. Reddy and W. H. Smith, Magnolia Publishing Inc. (1987).
34. A. Basseres, and Y. Pietrasanta, *Water Sci. Technol.* **24,** 229 (1991).
35. A. S. McAnally and L. D. Benefield, *J. Environ. Sci. Health* **A27**, 903 (1992).
36. B. D. Tripathi and S. C. Shukla, *Environ. Pollut.* **69**, 67 (1991).
37. H. Xu, B. Wang, Q. Yang and R. Liu, *Water Sci. Technol.* **26**, 1639 (1992).
38. B. G. S. Prasad, W. Madhavakrishna and Y. Nayudamma, in *Proc. Int. Conf. Water Hyacinth* ed. G. Thyagarajan, UNEP, Nairobi (1984).
39. M. Delgado, E. Guardiola and M. Bigeriego, *J. Environ. Sci. and Health Pt A Environ. Sci. and Engineering* **30**, 1423 (1995).
40. P. Y. Yang, H. Chen, N. Kongricharoern and C. Polprasert, *Water Sci. Technol.* **27**, 115 (1993).
41. A. J. Whitehead, K. V. Lo and N. R. Bulley, in *Aquatic Plants for Water Treatment and Resource Recovery* ed. K. R. Reddy and W. H. Smith, Magnolia Publishing Inc., Orlando (1987).
42. C. Simeon and M. Silhol, *Water Sci. Technol.* **19**, 1663 (1987).
43. D. M. Ogburn and N. J. Ogburn, *Aquaculture and Fisheries Management* **25**, 497 (1994).
44. V. Ngo and D. Hogen, in *Aquatic Plants for Water Treatment and Resource Recovery* ed. K. R. Reddy and W. H. Smith, Magnolia Publishing Inc., Orlando (1987).
45. A. S. Beling, W.G.A. Nissanka and K. Abeynayake. *J. Nat. Sci. Council Sri Lanka* **20,** 237 (1992).
46. C. K. John, in *Proc. Int. Conf. Water Hyacinth* ed. G. Thyagarajan, UNEP, Nairobi, 699 (1984).
47. L. P. D. Dewet, H. J. Schoonbee, J. Pretorius and L. M. Bezuidenhout, *Water SA* **16**, 28(1990).
48. A. K. Sen and N. G. Mondal, *Water Air & Soil Pollut.* **49**, 1 (1990).
49. D. K. Saxena, *Proc. Nat. Acad. Sci. India Sect. B (Biol. Sci.)* **65**, 61 (1995).

50. P. Y. Yang and H. Chen, *Bioresource Technol.* **49**, 129 (1994).
51. S.-y. Tang and X.-w. Lu, *Ecological Engineering.* **2**, 243 (1993).
52. R. Trivedy and V. R. Gudekar, *Water Sci. Technol.* **19**, 103 (1987).
53. Q. Dai, Y. Chen, Y. Pi, H. Zhang, G. Xu, X. Zhang and W. Dai, *Chinese J. Applied Ecol.* **2**, 159(1991).
54. R. K. Trivedy, P. K. Goel, V. R. Gudekar and M. G. Kirpekar, *Indian J. Environ. Protection* **3**, 106 (1983).
55. P. Singaram, *Indian J. Environ. Health.* **36**, 197 (1994).
56. P. Vajpayee, U. N. Rai, S. Sinha, R. D. Tripathi and P. Chandra, *Bull. Environ. Contam. & Toxicol.* **55**, 546 (1995).
57. C. L. R. Pinto, A. Caconia and M. M. Souza, *Water Sci. Technol.* **19**, 89 (1987).
58. J. P. Fett, J. Cambraia, M. A. Oliva and C. P. Jordao, *J. Plant Nutrition* **17**, 1219 (1994).
59. R. K. Srivasta, S. K. Gupta, K.D.P. Nigam and P. Vasudevan, *Water Research.* **28**, 1631 (1994).
60. S. K. Jain, P. Vasudevan and N. K. Jha, *Water Research* **24**, 177 (1990).
61. M. F. Zaranyika, F. Mutoko and H. Murahwa, *Sci. Total Environ.* **153**, 117 (1994).
62. O. Saltabas, and G. Akcin, *Toxicological & Environ. Chem.* **41,** 131 (1994).
63. C. Heaton, J. Frame and J. K. Hardy, in *Aquatic Plants for Water Treatment and Resource Recovery* ed. K. R. Reddy and W. H. Smith, Magnolia Publiching Inc. (1987).
64. G. Akcin, O. Saltabas and H. Afsar. *J. Environ. Sci. Health* **A29,** 2177 (1994).
65. W. H. Ornes, K. S. Sajwan, M. G. Dosskey and Adriano, *Water Air & Soil Pollut.* **57**, 53 (1991).
66. Y. Hao, A. L. Roach and G. J. Ramelow, *J. Environ. Sci. Health* **A28**, 2333 (1993)
67. X. Ding, J. Jiang, Y. Wang, W. Wang and B. Ru, *Environ. Pollut.* **84**, 93 (1994).
68. M. Coquery and P. M. Welbourn, *Water Research* **29**, 2094(1995).
69. M. Gupta, U. N. Rai, R. D. Tripathi and P. Chandra, *Chemosphere* **30**, 2011 (1995).
70. E. Grill. in Metal ion Homeostasis, *Molecular Biology and Chemistry* ed. D. H. Hamer and D. R. Winge, Alan R. Liss, Inc., New York (1989).
71. P. Guilizzoni, *Aquatic Bot.* **41**, 87 (1991).

72. B. C. Wolverton, *Proc. Aquaculture Systems for Wastewater Treatment,* USEPA 430/9-80-006 (1979).
73. K. K. Vrochinskiy, *Hydrobiol. J.* **6**, 103 (1970).
74. S. Roy and O. Hanninen, *Environ. Toxicol. & Chem.* **13**, 763(1994).
75. S. Roy and P. Lindstromseppa, S. Huuskonen and O. Hanninen. *Chemosphere* **30**, 1489 (1995).
76. M. H. Haggblom and L. Y. Young, *Applied and Environ. Microbiol.* **61**, 1546 (1995).
77. E. Grund, A. Schmitz, J. Fiedler and K. H. Gartemann in *Biotransformations , Microbial Degradation of Health Risk Compounds* ed. V. P. Singh, Elsevier Science, Amsterdam (1995.)
78. K. R. Reddy, E. M. D'Angelo and T.A. DeBusk, *J.Environ. Qual.* **19**, 261 (1990).
79. R. D. Martyn, D. A. Samuelson and T. E. Freeman, *J. Aquatic Plant Management* **21**, 49 (1983).
80. W. Larcher, *Physiological Plant Ecology.* 3rd edition, Springer-Verlag, Berlin (1995).
81. D. Elakovich St. *Biol. Plant.* **31**, 479 (1989).
82. S. Y. Yang, Z. W. Yu, W. H. Sun, B. W. Zhao, S. W. Yu, H. M. Wu, S. Y. Huang, H. Q. Zhou, K. Ma and X. F. Lao, *Acta Phytophysiologica Sinica* **18**, 399 (1992).
83. W. H. Sun, S. W. Yu, S. Y. Yang, P. W. Zhao, W. W. Yu, M. H. Wu, S. Y. Huanga and C. S. Tang, *Acta phytophysiologica Sinica* **19**, 92 (1993).
84. K. S. Clough, T. A. DeBusk and K. R. Reddy, in *Aquatic Plants for Water Treatment and Resource Recovery,* ed. K.R. Reddy and W.H. Smith, Magnolia Publishing Inc., Boca Raton. (1987).
85. K. R. Reddy and W.F. DeBuskm, *Econ. Bot.* **38**, 229 (1984).
86. K. K. Moorhead and K. R. Reddy, *Aquatic Bot.* **37**, 153 (1990).
87. W. T. Haller, D. L. Sutton and W. C. Barlowe, *Ecology* **55**, 891 (1974).
88. M.K.C. Sridhar and B. M. Sharma, *Experientia* **36**, 953 (1980).
89. K. R. Reddy and W.F. DeBusk, *J. Environ. Qual.* **14**, 459(1985).
90. G. Oron, *Agricultural Water Management* **26**, 27 (1994).
91. J. Vymazal, *Algae and element cycling in wetlands.* Lewis Publishers, Boca Raton (1995).
92. N. Boniardi, G. Vatta, R. Rota, G. Nono and S.Carra, *Chem. Engineering J. & Biochem. Engineering J.* **54,** B41 (1994).
93. O. Hammouda and M. S. Abdelhameed, *Folia Microbiologia* **39**, 420 (1994).

94. P. Hanczkowski, B. Szymczyk and M. Wawrzynski, *Animal Feed Sci. And Technol.* **52**, 339 (1995).
95. S. Kitoh, N. Shiomi and E. Uheda, *Aquatic Bot.* **46**, 129 (1993).
96. N. Shiomi and S. Kitoh, *J. Plant Nutrition* **10**, 1663 (1987).
97. R. Biljetina, V.J.Srivastava, D.P.Chynoweth and T.D.Hayes, in *Aquatic Plants for Water Treatment and Resource Recovery,* ed. K.R. Reddy and W.H. Smith, Magnolia Publishing Inc., Orlando. (1987).
98. M. Delgado, E. Guardiola and M. Bigeriego, *J. Environ. Sci. and Health Pt A Environ. Sci. & Engineering* **27**, 347 (1992).
99. J. C. Joyce, in *Aquatic weeds* ed. A. H. Pieterse and K. J. Murthy, Oxford University Press, Oxford (1990).
100. P. Lowilai, K. Kabata, C. Okamoto and M. Kikuchi, *Nippon Sochi Gakkai-Shi* **40**, 271 (1994).
101. C. Polprasert, N. Kongsricharoern and W. Kanjanaprapin, *Waste Management & Res.* **12**, 3 (1994).
102. N. R. Dhar, S. S. Singh, R. P.Chauhan and H. G. Sharma, *Proc. Nat. Acad. Sci. India* **40A**, 459 (1970).
103. B. Gopal, *Water Hyacinth.* Elsevier Science Publishers B. V. Amsterdam (1987).
104. T. Goswami and C. N. Saikia, *Bioresource Technol.* **50**, 235 (1994).

THREATENED WETLANDS AND THEIR RESTORATION

P. R. CHAUDHARI AND REKHA SARKAR

National Environmental Engineering Research Institute
Nehru Marg, Nagpur 440 020, India

1 Introduction

Wetland ecosystems are complex ecosystems of immense socio-economic and ecologic importance, and are regarded as the national wealth of significance. Wetlands play critical role in the maintenance and improvement of water quality of the natural water bodies due to natural processes of arresting the pollutants (organic and inorganic) discharged through direct releases, removal of sediments, production of oxygen, recycling of nutrients, and treatment of wastewater by removal of carbon, nitrogen, phosphorus, bacteria and viruses[1,2,3]. The wetlands also act as habitat for fish and shellfish, birds, mammals, reptiles and amphibians, many of which are of economic value in terms of subsistence and commercial fishing, hunting and trapping[4]. Non-consumptive uses of the wetlands, viz; bird watching, recreation, aesthetics, educational activities, and scientific research are also well recognised. Unfortunately, wetlands spread throughout the world are threatened or significantly altered due to pressures of enhanced human activities.

The present paper summarises the principles of wetland ecology and reviews in brief factors threating wetlands ecosystems,conservation and eco-management steps.

2 Definitions and Classifications of Wetlands

Wetland is a unique ecosystem which occupies transitional zone between permanently wet and generally dry environments and share characteristics of both these environments; yet this ecosystem can not be classified exclusively as either aquatic or terrestrial. Definition and classification of wetlands has been a problem, partly because of the enormous variety of wetland types and their highly dynamic character, and partly because of difficulties in defining their boundaries with precision as it is a complex ecosystem. Following definitions to explain and delineate wetlands have been put forth at various forums.

- Wetlands are submerged or water-saturated lands; both natural and man-made; permanent or temporary; with water that is static or flowing, fresh, brackish or salt, including areas of marine water; the depth of which at low tide does not exceed six metres (IUCN)[5].
- Wetlands are areas dominated by herbaceous macrophytes, the production of which takes place predominantly in the aerial environment above the water level while the plants are supplied with amounts of water that would be excessive for most other higher plants (IBP)[6].
- Part of the surrounding ecological structure and seral stages in the succession from open water to dry land and vice versa, occurring at sites situated as a rule between the highest and lowest water levels as long as the flooding or water logging of the soil is of substantial ecological significance (IBP)[6].
- Wetlands are lands transitional between terrestrial and aquatic systems where the water table is usually, at or, near the surface; or the land is covered by shallow water[7] (Cowardin *et al.*).

Wetlands exhibit enormous diversity according to their genesis, geographical location, water regime and chemistry, dominant plants, and soil or sediment characteristics. Many different types of wetlands may be found in close proximity forming not just different ecosystems, but wholly distinctive landscapes.

2.1 Classification

Essential features of wetland, by definition, include the ecologically wet systems such as tidal wetlands, fresh water wetlands, and artificially created wetlands. Three components of the ecosystem viz. vegetation, hydrology and soil type, play a vital role in defining the ecological processes and are associated with vital benefits to the wetlands.

Wetlands are broadly classified as (Figure 1)

- Natural
- Artificial or man-made & constructed wetlands

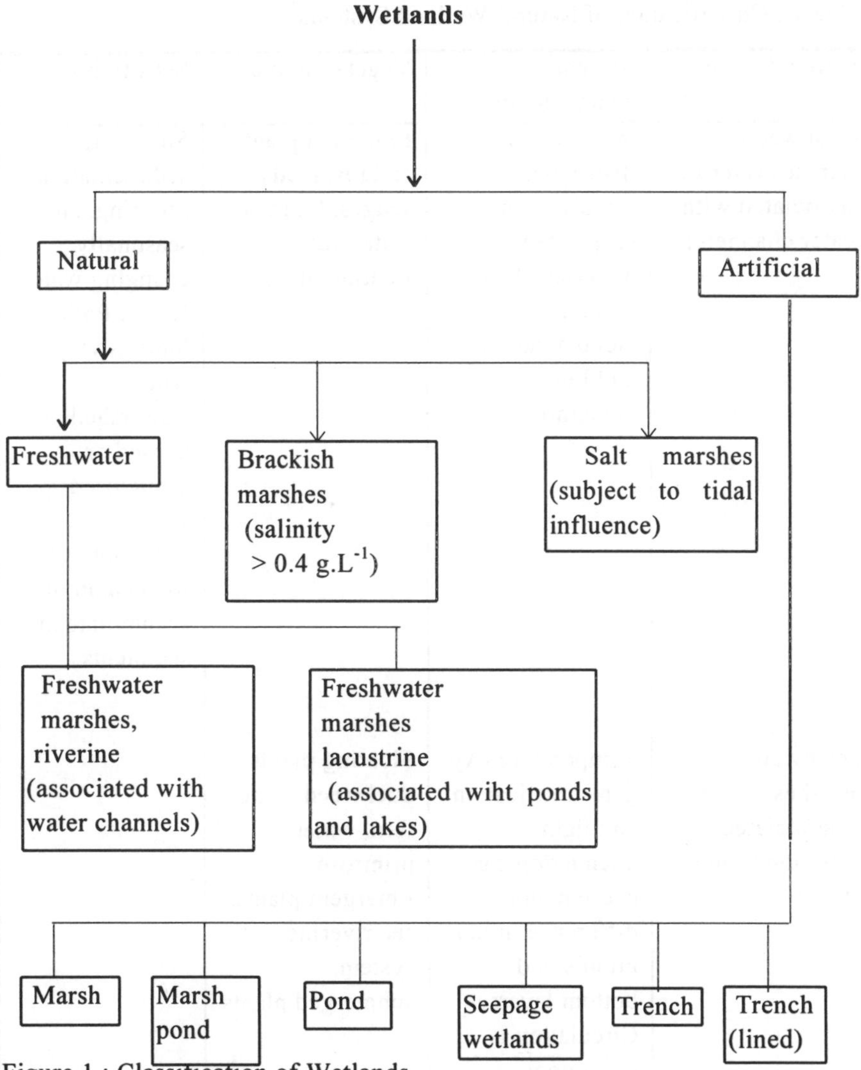

Figure 1 : Classification of Wetlands

2.2 Natural Wetlands

Natural wetlands are essentially classified on the basis of biological and habitat characteristics (Table1)[8].

Table 1 : Classification of Natural Wetland Systems*

Wetland system	System characteristics	Vegetation types	Sensitivities
Freshwater marshes:riverine (associated with water channels)	Water circulation distributes dissolved and suspended materials through system. Good aeration and light penetration	Emergent plants: cattails, reeds, sedges, bulrush, watercress; floating algae	Subject to sedimentation scouring, and seasonally changing water levels. Pollutant lodings vary with waterrshed. Closed or semiclosed systems. Pollutants enter food chain or accumulate in sediments
Freshwater marshes:lacustrine (associated with ponds and lakes)	Temperature/oxygen stratification and light attenuation can cause major difference in top, middle and bottom layers. Circulation is often poor	Floating plants : duckweed, water fern, water primrose, Emergent plants, see riverine system, submerged plants	
Freshwater marshes: palustrine (not confined by channels or adjacent to lakes)	Surface layer has thick and/or porous deposits with high organic contents. Marsh is fed by	Peat bogs, cypress, mangrove and papyrus swamps; vegetation types often specific to	Isolation from open water bodies (streams, rivers & lakes) limits water exchanged, forming

	subsurface seepage/high ground water.	geographical area	potential pollutant sink
Brackish marshes (Salinity > 0.4 ppt)	Marsh fed by seasonal surface flows and/or seepage; many also be subject to tidal influences; can experience salinity fluctuations	Emergent plants: sedges, bulrush, prckleweed, saltgrass, saltbush	Evporation can lead to salinities of 60-80ppt and concentration of pollutants
Salt marshes (suhject to tidal influence)	-Wetlands near streams -Lower wetlands, reversing flow -Lower wetlands, drained only at low tides -Upper wetlands, inundated only at high tides	Emergent plants: pickleweed, cordgrass, sedges, saltgrss	Salinity and sediment interactions can trap pollutants: however, low pH and oxidizing muds can re-release pollutants to system on a continuing basis

* Chan et al (1981)

They occur in a wide range of physical settings between terrestrial and aquatic environments. It is basically a complex ecosystem having detritus based food chain (Figure 2)[9]. US E.P.A.[10] has provided classification of natural wetlandsalong with system characteristics, vegetation types and sensitivities. All wetlands differ in characteristics, vegetation and sensitivities throughout the system.

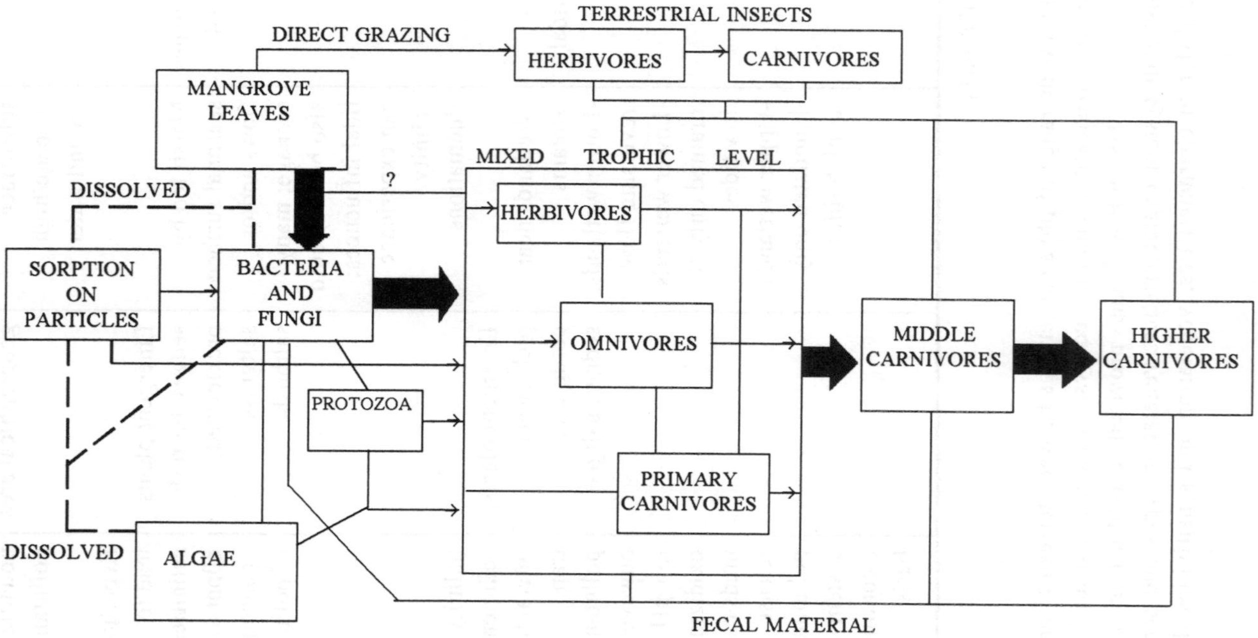

Figure 2 : Detritus food Chain in Wetland/Mangrove Vegetation Compartment model (Odum[12])

2.3 Artificial Wetlands

Artificial or man made, and constructed wetland systems are created by establishing wetland vegetation and requisite hydrologic conditions in locations where they previously existed. These constructed systems range from creation of marsh in a natural setting to intensive construction involving earth moving, grading, impermeable barriers, or erection of storages, viz. tanks, lakes, pools or trenches. The vegetation that is introduced or emerges as native from the constructrd wetland system will either be a monoculture or cultures similar to that found in natural wetlands. The types of artificial wetland systems are listed in Table 2[8]. Once constructed these systems will provide the same basic hydrologic and vegetative treatment functions as natural wetlands.

2.4 Values of Wetlands

Wetlands serve wide variety of functions including flood control, water purification, shoreline stabilisation and the control of erosion. They also support vast numbers of fish and other wildlife and numerous people depend on them for their livilihood. The value of the wetlands is summarised in Table 3.

Table 2 : Artificial Wetlands

Type	Description
Marsh	Areas with impervious to semipervious bottoms planted with various wetlands plants such as reeds or rushes
Marsh-pond	Marsh wetlands followed by ponds
Pond	Ponds with semipervious bottoms with embankments to contain or channel the applied water. Often, emergent wetland plants will be planted in clumps or mounds to form small subecosystems
Seepage wetlands	Wastewater irrigated fields overgrown with volunteer emergent wetland vegetation as a result of intermittent ponding and seepage of wastewater
Trench	Trenches or ditches planted with reeds or rushes. In some cases the trenches have been filled with peat
Trench (lined)	Trenches lined with an impervious membrane (lined) usually filled with gravel or sand and planted with reeds

* Chan et al.[8]

Table 3 : Summary of Wetland Values

Functions	Products	Attributes
Ground water recharge Ground water discharge Flood control/ erosion control Sediment/toxicant Nutrient retention Biomass export Strom protection/ wind break Water transport Recreation/tourisms	Forest resources Wildlife resource Fisheries Forage resources Agri-resources Water supply	Biodiversity Uniqueness to culture / heritage

3 Wetlands: The Global Scenario

A survey on the world's wetlands distribution was carried out by Finlayson and Moser[11]. A record of major global freshwater wetlands is appended in Table 4.[12]

3.1 Europe and the Mediterranean Basin

Europe and the Mediterranean regions are densely populated and have had such long histories of civilization and industrialisation that there are only a few entirely natural wetlands left in this area. Human interference has been less severe in parts of Icelands, the northern European Taiga and Tundra, but in most other regions the wetlands have either been destroyed or threatened. It has been documented that towards the end of the 1970's, 10 % of France wetter areas and 60 % those of the U.K. and the Netherlands had been drained[11]. Data for all the European and Mediterranean Ramsar sites reveal that only 58 of the 318 wetlands are definitely not threatened in some way.

Table 4 : Global freshwater wetlands areas

Region	Bogs	Fens	Swamps	Marshes	Flood plains	Lakes	Total
USSR	917	531	25	39	-	-	1,512
Europe	54	93	1	4	1	1	54
Near East	-	-	-	8	-	-	8
China	11	-	3	18	-	-	32
South East Asia	197	-	44	-	-	-	241
Australia /Newzealand	2	3	1	-	9	-	15
Africa	-	-	85	57	174	39	355
Alaska	?	250-400	?	?	?	?	(325)
Canada	673	531	14	44	-	6	1,268
USA**	13	-	80	40	95	-	228
Central America	-	-	15	2	1	-	18
South America	-	-	851	62	543	68	1,524
Total	1,867	1,483	1,130	274	823	114	5,691

* In 1,000 Kms ** Excluding Alaska

Source : Aselmann, and Crutzen[12].

3.2 North America

Canada is estimated to hold 24 % of the world's wetlands, occupying over 1.27 million Km^2. The original wetland area of the conterminous USA (excluding Alaska and Hawaii) may have been around 890,000 Km^2 of which only 47 % remain[13]. There were a further 690,000 Km^2 in Alaska and Hawaii, with only a very small percentage in the later state. Overall, it is estimated that around 1.11 million Km^2 of wetlands remain in the whole of the USA[13].

3.3 Latin America and the Caribbean

The wetlands of south America have been identified as those of the Atlantic-Caribbean lowlands to the north and east of the Andes. The Chilean Fjordland in the Pacific lowlands includes around 55,000 Km^2 of wetlands. The largest wetland in the Andes region is the freshwater lake, lake Tota, in Colombia and Chilean lake district covering about 3,000 Km^2 of wetland. In the lowlands of the Atlantic-Caribbean region, the delta of the River Orinoco covers an area of around 30.000 Km^2, while that of the River Amazon covers about 35,000 Km^2. On the other site, the Pantanal covering an area of about 200,000 Km^2 is one of the largest flood plains in the world.

Many of the Caribbean islands have important wetlands, mostly coastal lagoons, mangrove swamps and intertidal mudflats, but there are also some freshwater lakes in old volcanic craters. The Usumacinta delta is the most extensive wetland on the Gulf Coast of Mexico, covering around 10,000 Km^2.

3.4 Africa

The wetlands cover one percent of Africa's total surface area (atleast 3,45,000 Km^2). In Equatorial Africa, the largest wetlands system are : the Zaire swamps (covering 80,000 Km^2), the Sudd in the upper Nile (over 50,000 Km^2) and the wetlands of the Lake Victoria Basin (about 50,000 Km^2). The flood plain of the Niger and Zambezi Rivers, the Chad basin (around 20,000 Km^2) and the Okavango delta (16,000 Km^2) are also some of the major wetland areas. About 12,000 Km^2 of wetland also exist in southern Africa.

3.5 Asia and the Middle East

It has been estimated that there are about 8,30,000 Km^2 of peat bogs and swamps in the USSR and about 9,00,000 Km^2 of marshy ground subject to seasonal flooding (Finlayson and Moser)[11].

In the Middle East, the most extensive wetlands occur in Iraq, where the Tigris and Euphrates rivers create a vast complex of shallow lakes and marshes covering about 15,000 Km^2. The important wetlands of Asia are shown in Table 4 and 5[12,14].

Table 5 Important Wetlands of Asia

Country	Number of sites	Area of sites
Bangladesh	12	67,700
Bhutan	5	85
Brunei	3	1,380
China	192	1,63,000
Hong Kong	3	119
India	93	54,700
Indonesia	137	87,800
Japan	85	4,750
Combodia	4	36,500
Koren, Dem People's Rep	15	3,222
Korea, Dep	21	1,070
Laos	4	2,220
Malaysia	37	31,200
Mongolia	30	15,500
Myanmer	18	54,900
Nepal	17	356
Pakisthan	48	8,580
Papua New Guiner	33	1,01,000
Philippines	63	14,100
Singapore	7	2
Srilanka	41	2,740
Taiwan	12	84
Thailand	42	25,100
Vietnam	25	58,100
Total	947	7,34,200

Source : Scott & Poole[14]

3.6 Wetlands: Indian Scenario

India is ecologically rich in biological diversity, owing to its location (8-38°N and 60-97.5°E) and extensive habitat variations. On an average, this region harbours over 45,000 species of plants and 65,000 species of animals.

Wetlands in India is the abode of native flora and fauna and a roosting place for migratory birds. These exhibit significant ecological diversity, primarily because of variability in climatic conditions and changing topography.[15] Many wetland areas are getting reduced due to their diversion to aquaculture (prawn culture and fish culture) and agriculture activities, industry and settlement. Disposal of industrial effluents, municipal wastewater and sedimentation due to ecological degradation in catchment areas have resulted in adverse impact on the wetlands.[16,17]

Ministry of Environment and Forests (MEF)[18] brought out an Inventory on Wetlands in India[8] in which the wetlands in India have been listed along with the list of research projects sponsored, to prepare management plans, for the conservation of 13 wetlands in India. This has been an initial step towards protection of wetlands.

Maximum number of natural wetlands are in Assam (1394) followed by Andhra Pradesh (219), Uttar Pradesh (125), Bihar (65), West Bengal (54), Maharashtra (49). Large natural wetland areas have been recorded in Gujarat, West Bengal, Bihar, Orissa and Andhra Pradesh. Areawise largest man-made wetland area is recorded in Karnataka followed by those in Andhra Pradesh, Maharashtra, Uttar Pradesh, Kerala, Tamil Nadu, Madhya Pradesh, Orissa, Gujarat and Rajasthan. Important wetlands in India are Kolleru lake, Pulicat lake (Andhra Pradesh), Dal lake (Kashmir). According to the report from Ministry of Environment & Forests, the total number of natural wetlands in India (including wetlands less than 100 ha) are 2167 (Table 6)[18] with an area of 14,50,871 ha and that of artificial wetlands are 65,254 covering an area of 2,58,266 ha.

4 Threats to the Wetlands

The wetland ecosystems are seriously threatened due to pressures of rapid urbanisation, agriculture and industrial development, increasing levels of pollution, over exploitation, and hydrological alterations. Over one million hectares of mangroves annually are being destroyed worldwide. Wetlands and mangroves are also undergoing rapid destruction in India (MEF,1990)[18].

Indiscriminate conversion of wetlands for agriculture and human habitats, sewage disposal, siltatation, eutrophication, invasion of weeds, converting parts of these wetlands into swamps, marshes and drylands and demographic.

pressures have been identified as the main causes of the degradation of wetlands throughout the world (Table 7).[14]

Table 6 State/Union Territory-Wise Distribution of Wetlands in India (including wetlands of less than 100 ha)

State	Natural		Man-Made	
	No.	Area(ha)	No.	Area(ha)
(1)	(2)	(3)	(4)	(5)
1. Andhra Pradesh	219	1,00,457	19,020	4,25,892
2. Arunachal Pradesh	2	20,200	NA	NA
3. Assam	1394			
4. Bihar	62	2,24,788	33	48,607
5. Goa	3	12,360	NA	NA
6. Gujarat	22	3,94,627	57	1,29,660
7. Haryana	14	2,691	4	1,079
8. Himachal Pradesh	5	702	3	19,165
9. Jammu & Kashmir	18	7,227	NA	21,880
10. Karnataka	10	3,320	22,758	5,39,195
11. Kerala	32	24,329	2,121	2,10,579
12. Madhya Pradesh	8	324	53	1,87,818
13. Maharashtra	49	21,675	1,004	2,79,025
14. Manipur	5	26,600	NA	NA
15. Meghalaya	2	NA	NA	NA
16. Nagaland	2	210	NA	NA
17. Orissa	20	1,37,022	36	1,48,454
18. Punjab	33	17,085	6	5,391
19. Rajasthan	9	14,027	85	1,00,217
20. Sikkim	42	1,101	2	3.5
21. Tamil Nadu	31	58,868	20,030	2,01,132
22. Tripura	3	575	1	4,833
23. Uttar Pradesh	125	12,832	28	2,12,470
24. West Bengal	54	2,91,963	9	52,564
Total	2,164	14,49,338	65,250	25,87,965
Union Territories				
1. Chandigarh	NA	NA	1	170
2. Pondicherry	3	1,533	3	1,131
Grand Total	2,167	14,50,871	65,254	25,8,266

* Ministry of Environment and Forests (1990)[18]

Table 7 Major Threats to Wetlands of the World

Country	Threat(s)	Incidence(% of sites)
Latin America and Caribbean	Pollution	31.0
	Hunting & associated disturbance	30.5
	Drainage for agriculture	19.0
	Disturbance for recreation	11.5
	Reclamation for urban and industrial development	10.5
	Forestry activities	10.0
	Fishing and associated disturbance	10.0
Asian Countries	Hunting and associated disturbance	32.0
	Human settlement/ enchroachment	27.0
	Drainage for agriculture	23.0
	Pollution	20.0
	Fishing & associated disturbance	19.0
	Commercial logging/ forestry	17.0
	Wood cutting for domestic use	16.0
	Degradation of water shed/soil erosion/siltation	15.0
	Conversion to aqua-culture ponds or salt pans	11.0
	Diversion of water supply	9.0
	Overgrazing by domestic stock	9.0

Source : Scott and Poole[14].

Severity of threats to wetlands of international importance are depicted in Table 8.[14]

Table 8 Severity of Threats to Wetlands of Intrnational Importance

Country	Number of Sites Known	Degree of Threat None	Low	Mode-rate	High	% Sites with Moderate to High Threat
Bangaladesh	11	1	1	5	4	82
Bhutan	5	3	-	1	1	40
Brunei	3	-	2	1	-	33
China	105	30	34	36	5	39
Hongkong	3	1	-	2	-	67
India	88	4	44	22	18	45
Indonesia	129	1	54	66	8	57
Japan	38	8	11	17	2	50
Cambodia	3	-	1	2	-	67
Korea, DPR	5	5	-	-	-	0
Korea, Rep	19	5	3	6	5	58
Laos	3	-	1	2	-	67
Malaysia	37	-	5	22	10	86
Mangolia	30	23	5	2	-	7
Myanmar	16	-	7	8	1	56
Nepal	14	2	7	4	1	36
Pakistan	42	1	20	15	6	50
Papua New Guinea	27	14	8	4	-	15
Philippines	49	2	13	24	10	69
Singapore	6	-	2	3	1	67
Srilanka	31	2	8	13	8	
Taiwan	12	1	4	5	2	58
Thailand	36	1	18	14	3	47
Vietman	23	3	14	4	2	26
Total	734	107	262	278	87	50

Source : Scott and Poolle[14].

Due to lack of adequate public awareness, wetlands are considered to be wastelands providing habitats for breeding and growth of mosquitoes and molluscs which serve as vectors for number for deadly diseases like malaria, filariasis and schistosomiasis. Malaria epidemics caused by Anopheles mosquitoes have been, and continue to be, major cause of human suffering and mortality in Indian subcontinent. Large areas of seasonal wetlands are therefore considered undesirable. These wetlands are drained, filled with soil or solid waste, and are put to use for greater economic returns. On the other hand river courses are regulated, diverted, and dammed to eliminate their flood plains.

Conversion of wetlands for urbanisation, industrialisation, agriculture, and aquaculture (fish and shrimp culture) are also very common phenomena throughout the world. In coastal areas of India, rapidly proliferating shrimp culture activity is also responsible for the destruction of mangroves. Intertidal areas are generally chosen to take advantage of natural tidal flushing. Recent studies have shown 16% decrease in mangrove extent and 27% increase in heavy urban and industrial development of coastal zones in Ecuador. This is causing an influx of pollutants into coastal and esturine areas. Scott and Poole[14] have listed countries with most seriously threatened wetlands in Asia (Table 9).

Table 9 The most Seriously Threatened Wetlands in Asia

Bangladesh	13. Xi Jiang (Pearl River) Delta*
1. Chalan Beel*	14. Tuosu Hu (Kurlyk Nor) and Kuerhleiko Hu
2. Haor Basin of Sylhet and Eastern Mymensingh	**India**
3. Dubriar Haor*	15. Dal Lake
4. The Sunderbans	16. Wular Lake
5. Wetlands in Pablakhali Wildlife Sanctuary	17. Harike Lake
6. Chokoria Sunderbans*	18. Jheels Near Haidergarh*
Bhutan	19. Dahar and Saju (Soj) Jheels
7. Boomtheng Valley	20. Southern Gulf of Kutch
Burma	21. Gulf of Khambhat
8. Irrawaddy Delta	22. Khabartal
China	23. Dipor (Deepar) Bheel
9. Yan cheng Marshes	24. Logtak Lake
10. Shijiu Hu	25. Salt Lake Swamp*
11. Shengjin Hu and the lower Yangtze Lakes	26. Sunderbans
12. Shengjin Hu	27. Chilka Lake
	28. Kolleru Lake
	29. Estuaries of Karnataka Coast

30. Kaliveli Tank & Yedayanthittu Estuary
31. Cochin Backwaters
32. Wetlands of Andaman & Nicobar Islands

Indonesia

33. Banyuasin Musi River Delta
34. Muara Cimanuk*
35. Sukolilo
36. Cilacap and Segare Anaken
37. Danau Bankau & other Swamps in the Barito Basin*
38. Banau Sentarum
39. Wetlands in Manusela Proposed National Park
40. Wasur and Rawa Biru

Japan

41. Shonai-Fujimae Tidal Flats & Inner Isee Bay
42. Lake Shinji Lake Nakaumi

Korea

43. South Kanghwa and North Yongjong Mudflats
44. Mudflats of South Yongjong & adjacent Islands
45. Namyang Bay
46. Asan Bay
47. Kum Mankyung & Tangjin Estuaries

Malaysia

48. Sedili Kecil Swamp
49. Kalang Islands : Pulau Kelam*

Eastern Mymensing

50. Kapar Forest reserves
51. North Selangor Swamp Forest
52. Marintaman Mengalong*
53. Tempasuk Plain
54. Lawas Mangroves
55. Trusar-Sundar Mangroves
56. Limbang Mangroves
57. Maludam Swamp Forest
58.Sarawak mangrove Forest

Nepal

59. Begnas Tal*

Pakisthan

60. Khabbaki Lake*
61. Siranda Lake·
62. Hawekes Bay/ Sandspit Beaches and adjacent Creeks
63. Clifton Beach
64. Korangi and Gharo Creeks
65. The Outer Indus Delta

Philippines

66. Pangasiunan Wetlands
67. Manila Bay*
68. Laguna de Bay
69. Tayabas Bay including Plagbilao Bay
70. Impanga Coast
71. Mactan, Kalawisean and Cansaga Bays*
72. Agusan Marsh
73. Lake Leonard
74. Davao Gulf
75. Liguasan Marsh

Singapore

76. Serangoan Estuary*

Sri Lanka

77. Thandamannar Lagoon*
78. Chundikkulam Lagoon
79. Chalai Lagoon*
80. Periyakarachchi and

Sinnakarachchi Lagoon*
81. Mahaweli Ganga Flood-plain System
82. Maha Lewaya and Karagan Lewaya
83. Lunama Kalapuwa and Kalametiya Kalapuwa*
84. Bellanwilla - Attidiya Marshes

Taiwan, R.O.C.

85. Tatu Estuary
86. Tungshih (Tan-shou) Mangroves*

Thailand

87. Gulf of Thailand
88. Pak Phanang Estuary
89. Pa Phru

Socialist Republic of Vietnam

90. Red River Delta
91. Red River Estuary
92. Mekong delta
93. Nam can Mangrove Forest.

Source : Scott, and Poole[14].

* Sites are considered to be already to degraded to merit any special conservation efforts.

Wetland food chain is essentially a detritus based foodchain which is affected due to influx of toxicants, leading to reduction in the productivity of wetlands. Problem of influx of pollutants to natural water bodies due to encroachment on wetlands has acquired a global dimension in the recent past.

5 Wetland Restoration : Experiences

5.1 Management

Comprehensive management plans for wetland reserves, to redress effects of human impacts, are accepted as essential tools throughout the world.[19] The integration of wetlands into regional and national planning has also become established with schemes for water resources and river basin development in some countries.[20-23]

Probably the most exciting aspect of management in Europe and Mediterranean at present is wetland restoration. Restoration schemes have already been completed at three Ramsar sites. These restoration schemes are related to resumption of water supply and maintenance of natural depth of water in wetlands.

Lake Fetzara in Algeria was drained in the 1930s, but has recently been restored as a lake. Local hydraulic engineers used it to store floodwater in an emergency and found that they had created a valuable source of irrigation water and grazing land. In the Netherlands, lake restoration scheme seeks to remove some of the summer dykes from the river to allow flooding between the permanent winter dykes.

The Swedish Government has decided that the maximum efforts must be put into managing existing wetlands rather than restoring those which have been degraded. Even so, there are at present four state restoration projects under consideration.

In Denmark, 40 % of floodplains of largest river, the Skjern, were converted into arable land in 1969. But farming activities in the delta and upstream areas resulted in excessive eutrophication, causing serious depletion of flora and fauna. However in 1990, the Government began implementation of a restoration project that will reinstate natural nutrient filtration processes, thus allowing the ecosystem to recover its former productivity.

5.2 Conservation

Within Europe and the Mediterranean region, most of the countries are Contracting Parties to the Ramsar Convention. More than 300 wetlands have so far been listed in the region and some countries, such as Italy and the United Kingdom, have listed many sites of international importance. The largest Ramsar wetlands in the region are the Danube Delta in Romania and the Wadden Sea, where Denmark, Germany and the Netherlands have together designated five sites covering some 6,00,000 hectares.[24]

European countries have always played a major role in the Convention. Today, many of these participating countries have laid down policies for assisting developing countries in conservation of wetlands to achieve sustainable development.

The European community does not have a wetlands policy as such, although certain conservation objectives are set out by various directives and these must be incorporated in the national law of member states. Perhaps the European Community's greatest contributions to wetlands conservation can be identified through their policy of removal of direct and indirect subsidies for wetland destruction, included in the Common Agricultural Policy and Structural Funds Programme.

Multi-lateral and bi-lateral banks and international development agencies have, in the past, promoted many projects which have had adverse impacts on wetlands. However, the work of the United Nations 'Brundtland' Commmission on Environment and Development has promoted the idea of sustainable development, and the international funding agencies have grown to appreciate the importance of environmental considerations, including wetlands. Protecting the enviroment is a central concern of the world bank and the European Investment Bank. Almost all countries have laws which provide for some element of wetland protection.

Strong non-governmental organisations (NGOs), with secure and independent funding from a large and well-informed membership base, are powerful forces for conservation. Such groups lobby for action, mount campaigns with strong support from their members, undertake independent research, promote awareness and information activities, buy or manage reserves (or both), and assist similar organisations in other countries.

Many of the wetlands in Europe and the Mediterranean have been either destroyed or degraded, and a majority of those that remain are under threat. However, the future of wetlands in Europe is probably quite optimistic because of the social changes that have brought public pressure for "greener" governmental policies with regard to agriculture and industry.

However, the rapidly increasing populations of the east and south of Mediterranean, and the existing environmental problems of many of those countries, suggest that it will be more difficult to conserve the wetlands jewels which still remain in this region.

In India, a major step was taken by enactment of Indian Wildlife (Protection) Act in 1972 with a view to conserve flora and fauna of the country. Such efforts were supplemented by Man and Biosphere Programme (MAB). Indian MAB committee in 1979 has identified a network of 13 representative ecosystems (some containing wetland ecosystems) to be protected as Biosphere Reserves.

A National Wetland committee has been constituted under Ministry of Environment & Forests, New Delhi under seventh plan for Protection of Wetlands, with a view to advise the Government on the policy and measures to be taken and selection of wetlands for conservation and management. The objective was that the practices evolved could be replicated elsewhere in the country. An action plan for wetland ecosystems was drawn at in the meeting of the National Wetland Committee which met in March 1988.

5.3 Environmental Management Plan

A stretagic approach for Integrated Environmental Management Plan (EMP) for wetland restoration has been formulated by NEERI[25], considering the experiences in relation to wetland restoration in the world[26] and in India particular[27,28].

Formulation of Environmental Management Plan specifically aims at the restoration of diversity of wetlands. Methodologies to mitigate the adverse impacts on wetlands, ecological restoration strategy and social awareness Programme through training and public participation in management processes are the vital issues which need to be addressed through the EMP (Figure 3). Planning of strategies and programmes for ecological restoration and mitigation of impacts of wetlands are enumerated hereunder

6 Ecological Restoration

6.1 Land Use Planning

Considering the data collected through the studies on distribution and quality of forests, land forms, drainage pattern in wetland area, current pattern and other environmental factors which sustain or contain the mangrove ecosystem, mangroves zones need be outlined for the following activities in coordination with national socio-economic development and coastal development plans[29,30].

Conservation zones : Protection of natural and relatively undisturbed mangrove ecosystem for various benefits.

Management zones : Management for sustained yield for timber production through silvicultural system needs to be carried out to meet the local requirements to minimise the adverse impact on wetlands.

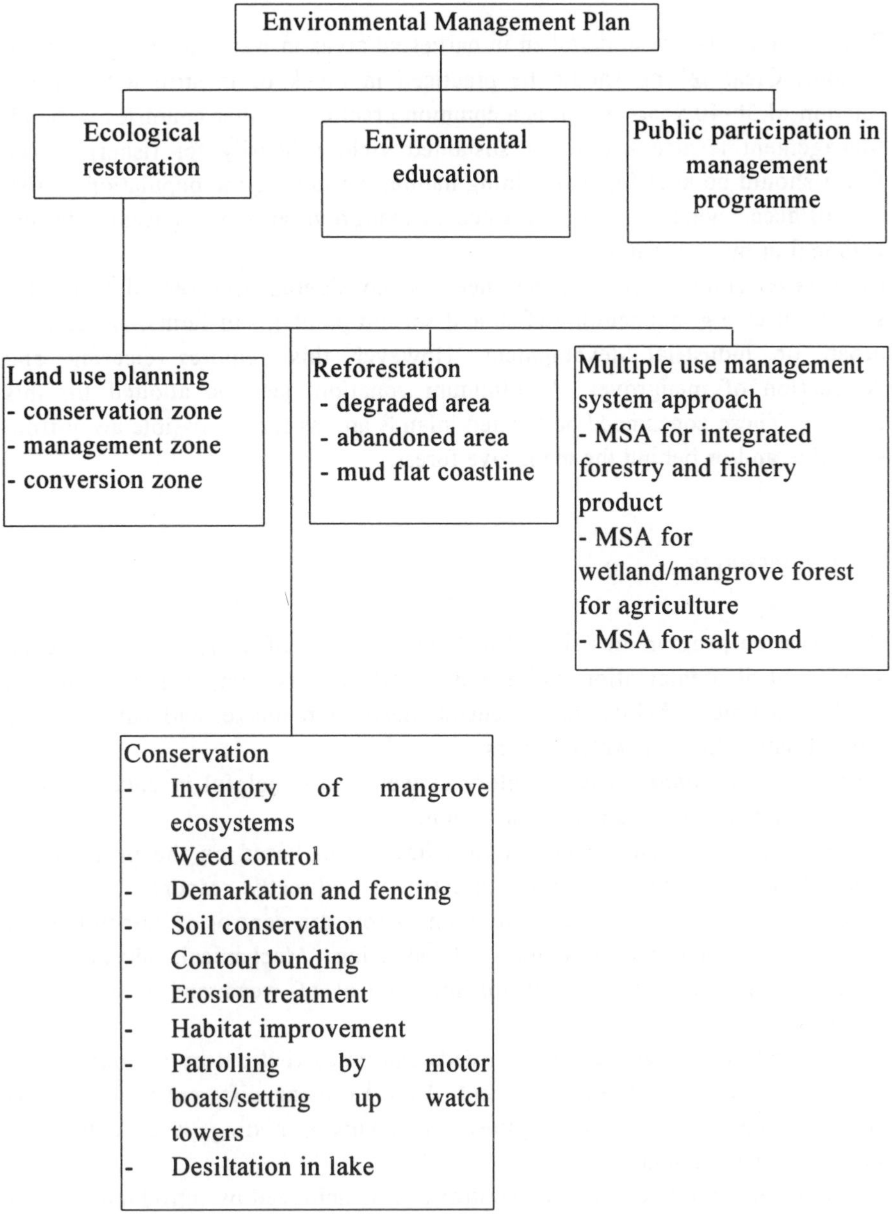

Figure 3 : Environmental Management Plan

Planting may also be undertaken in harvested areas in wetlands if regeneration is poor. Clear felling should be practiced in block or in strip with cutting rotation of 30-40 years which is a common practice in other countries wetland. Management is also useful for sustained yield primarily for fisheries. The forest should be kept for maintaining habitat, sustaining the population of fish & crustacean which can be harvested in mangrove area & adjacent esturine, lagoonal or marine waters.

Conversion zones : These zones need be considered as reserved zones for specific uses e.g. aquaculture (fish and shrimp ponds), salt farms, agriculture, urban or industrial development. However this requires clearing and destruction of mangroves. A minimum zonation need be allotted for this purpose. These zones need be located inlands and as far as possible away from shoreline and/or behind the mangrove forest.

6.2 Conservation

Conservation programmes should include inventory of mangrove ecosystems, weed control, demarcation and fencing, soil conservation, contour bunding, erosion treatment, habitat improvement, desiltation intake, and patrolling by motor boats/setting up watch towers.

Inventory of wetlands: The inventory of wetlands is helpful in understanding their ecological role in a particular region.

Weed Control : Aquatic weed control have been based on the principle of controlling a variety of factors directly or indirectly responsible for the stimulation of aquatic growth. The main factors are control of autochthonous sources & allochthonous sources of pollution. Mechanical, physical and biological methods are helpful for the control of weed and weed related problems.

Control of submerged aquatic weeds can be carried out by various simple tools. Physical methods of weed control are by using different tools such as machines, chaining or chain scythes, chain knives, netting, weed cutter and special harvesting boats.

Biological methods of weed control can be achieved by introducing insect pests and weedies. Biological control of noxious weed may be successful after economic uses of fisheries, irrigation, navigation etc. in wetlands.

Demarcation and fencing : Entry as well as encroachment of wetland by public can be prevented by making distinct fencing after proper demarcation. As the people are unable to enter the demarcated area, quality as well as resources will not be disturbed.

Soil conservation : Fertility of wetland soil is affected due to diversion of sediment as a result of construction of dams, channeling of streams etc. Soil conservation technique should be applied to restore the soil fertility.

Contour bunding : In order to prevent excessive loss of water because of natural drainage, contour bunding can be implemented around the wetlands at suitable places. This will ensure the adequate retention of water in the wetlands which is important for the survival of marshy plants.

Erosion treatment : In the degraded wetland areas the soil erosion is a common factor. To prevent erosion of soil in this area, planting of helophytes should be done which will be useful in preventing soil erosion.

Desiltation in lakes : The incoming water of the wetland may contain excessive amount of suspended matter because of human activities such as mining, afforestation, construction of roads etc. This will enhance the siltation in the lake. Proper controlled measures need to be adopted to reduce the presence of suspended matter in the influents of the wetland.

Habitat improvement : Wetland habitat alterations for control of disease/parasite vectors which must be mediated by the ecological needs of desirable wetland consumers, deserves great attention. By proper protection through physical, chemical and biological ways, habitat of the ecological flora and fauna should be improved. Improvement of habitat will lead to creation of stable ecosystem.

Patrolling by motor boats/setting up watch towers : Monitoring of wetland is most essential from the view point of preservation and conservation. Monitoring of different anthropogenic pressures on wetland need to be taken care of regularly. This monitoring can be carried out by installing watch towers as well as patrolling the wetland area by motor boats. Patrolling will be helpful in maintaining the safety and threshold criteria with reference to size, shape and location.

6.3 Reforestation

Reforestation needs to be implemented on degraded areas, and abandoned areas (particularly the areas being left after mining) left after fish or shrimp farming, and on mud-flat coastline. The important factors for success are tidal pattern, soil conditions, enemies and species itself. The areas to be planted should be flooded by sea water. Suitable species of wetland should be planted considering the type of environment.

6.4 Multiple use management system approach

Following multiple sustained use of the wetlands is preferred rather than conversion for single use

- Management system for integrated forestry and fishery product
- Management system for wetland/mangrove forest for agriculture
- Management system for salt pan

6.5 Environmental education

One method of conserving the wetland system is by creating general public awareness. Encouragement of eco-tourism is the other method for promoting scientific knowledge and generating revenue. Following issues need to be addressed to achieve wetland conservation through public awareness.

Public authorities and bodies responsible for management of wetland shall promote and facilitate by all appropriate methods, public access to wetlands and interpretation of their values.

For proper public awareness radio, television, magazine, journal, news papers, drama, road procession, cultural function related to eco-activities will be most effective. Besides course structure in school, university, research institutes should have better chance of public awareness.

6.6 Public participation in management programme

The real success or achievement of wetland conservation is dependent on public participation. Public meetings have facilitated public input during last few decades. Public participation for wider application of the wise use concept through adoption of nation wise use of policies with appropriate legislative support are important.

Sustainable management of wetlands depends on local, national and regional levels, bringing together people from the many disciplines required for an effective integrated approach to management.

Summary

Wetlands are unique ecosystems which occupy transitional zones between permanantly wet and generally dry environments sharing the characteristics of both these environments. Wetlands are recognised as complex ecosystems of immense value from ecologic and socio-economic points of view, and are considered to be national wealth of significance. Globally, wetlands are threatened or significantly altered due to pressures from enhanced human population, over-exploitation and environmental pollution.

Wetlands have been, till now, considered as wastelands. Their ecological and economical role is recently recognised. Efforts are being made on global level to protect and conserve wetland ecosystems. However, the task is difficult due to lack of systematic baseline studies, quantification of impacts and mitigation measures.

To avoid further losses of these unique ecosystems; scientific investigations are widely undertaken on wetland inventories; identification and quantification of existing pollution sources; establishment of threshold criteria; utilisation of wetlands for recovery of valuable natural resources through aquaculture; storage of water for irrigation and drinking; and biogeochemical cycling. Detailed assessment of wetland dynamics will lead to the development of baseline status which will be helpful to plan strategies for wetland conservation, restoration and sustainable management programmes. Ecomanagement of the wetlands may be carried out through the understanding of their structure and function, habitat alterations, consumer inventories, and land-use pattern in entire catchment area.

Present paper summarises the principles of wetland ecology, factors threatening wetland ecosystems, their conservation, and ecomanagement strategies.

Acknowledgement

The authors are thankful to Director, NEERI, Nagpur for the permission to present this paper in the International Conference to be held in Malaysia.

References

1. K.G. Boto, and Patrick Jr., in *Proceedings of the National Symposium on* Wetland. eds P.E. 1.Greeson, J.R. Clark and J. E. Clark, (1978).
2. R. G. Brown, Water Resources Investigation Report, United States Geological Survey, St. Paul, Minnesota (1985).
3. R.H. Kadlec, and D.L. Tilton, *J. Env. Qual.,* **8**, 328 (1979).
4. M.W. Weller, *Freshwater Marshes: Ecology and Wildlife Management.* University of Minnesota Press, Minneapolis (1981).
5. IUCN, *Spl. Supply IUCN Bull.,* **2,** 1 (1971).
6. IBP Fifth General Assembly (1972).
7. L.M. Cowardian, V. Carter, F.C. Golet, and E.T. LaRoc, Classification of wetland and deep water habitats of the United States, US Fish and WL Serv., Washington (1979).
8. E. Chan, T.A. Bursztynsky, N.N. Hantzsche, and Y.J. Litwin, The use of wetlands for water pollution control. U.S. EPA Grant No. R806357, Office of Research and Development, Cincinnati, Ohio (1981).
9. W.E. Odum, Dissertation. Univ. of Miami, (1970).
10. U.S. Environmental Protection Agency (US EPA) The lake reservoir restoration guidance manual. EPA-44075-88-002, (1988).
11. M. Finlayson, and M. Moser, eds. *Wetlands. Facts on the File Limited,* Oxford (1991).
12. I. Aselmann, and P.J. Crutzen, *J. Atmospheric Chem.***8**,307(1989).
13. T.E. Dahl, US Department of Interior, Fish and Wildlife Service, Washington D.C. (1990).
14. D.A. Scott, and C.M. Poole, A status overview of Asian Wetlands. No.53, AWB, Kula Lumpur, Malaysia (1989).
15. B. Gopal, Proc. Internat. Sci. Workshop on Ecosystem Dynamics in Freshwater Wetlands (1982).
16. B. Gopal, in *Aquatic Weeds.* eds. A.H. Pieterse and K.J. Murpjy Oxford University Press (1988).
17. B. Gopal, *Water Quality Bulletin,* **13**, 3 (1988).

18. Ministry of Environment and Forests. Conservation of mangroves in India. Govt. of India, New Delhi (1990).
19. J.B. Millar, *Can. J. Bot.,* **51,** 1443 (1973)..
20. J.A. Kadlec, *Ecology,* **43,** 267 (1962).
21. J.A. Kadlec, and W.A. Wentz, National Technical Information Service, Springfield (1974).
22. G.E. Hutchinson, *Limnological Botany.* Academic Press, New York, (1975).
23. L.H. Fredrickson, and T.S. Taylor, U.S. Department of the Interior, Fish and Wildlife Service, Res ource Publication, 148, Washington, (1982).
24. World Conservation Monitoring Centre *Global Biodiversity,* Chapman & Hall Publication, London (1992).
25. NEERI, Environmental Management for Prawn Farming Activities MPEDA, Kochi, (1995).
26. L.H. Suring, and M.D. Knighton, U.S. Department of Agriculture, Forest Service, General Technical Report NC-100, Washington, (1988).
27. NEERI, Aquatic Baseline Study for Thermal Power Station near Dahanu, BSES, Bombay, (1993).
28. NEERI, Management and control of Aquatic Weeds in Kanjali Lake, District Kapurthala (Punjab) (1993).
29. A.G. Van-der Valk, C.B. Davis, J.L. Baker, and C.E. Beer, in *Proceedings of the National Symposium on Wetland.* eds. P.E. Greeson, J.R. Clark and J.E. Clark (1978).
30. D.H.N. Spence, *Adv. Ecological Res.,* **15**, 37 (1982).

BIORESTORATION OF METAL DAMAGED FRESHWATER ECOSYSTEMS :

CURRENT STATUS AND FUTURE PERSPECTIVES

SUBHASHREE PRADHAN, SARITA SINGH AND L. C. RAI

Laboratory of Algal Biology,
Department of Botany,
Banaras Hindu University,
Varanasi- 221 005, India

1 Introduction

Rapid industrialization, urbanisation, increasing population, intensive agriculture, fast depletion of nonrenewable resource and stratospheric ozone shield that protects the earth from harmful UV-B radiation are some of the factors which pose a potential threat to the quality of life. High demand for various synthetic chemicals both in developed and developing nations has accelerated the rate of release of toxic contaminants into the environment thereby polluting our aquatic and terrestrial ecosystem. Of the various toxicants, heavy metals assume special importance as environmental pollutants after the incidents of Minamata Bay and Jintsu-river where fishermen died or became paralyzed due to mercury and cadmium poisoning. Increase in concentration of toxic metals into the environment has consequently augmented the chances of human exposure to these metals through ingestion, inhalation or skin contact.

Water is essential for sustenance of living beings as well as developmental needs, may it be irrigation, hydropower, industries, municipal supplies, fisheries or recreation etc. Despite the singnificane of freshwater bodies for human beings, no worth while attempts have been made to avoid their abuse. Nevertheless, freshwater ecosystems are subjected to such anthropogenic metal input sources as Cd from electroplating and nuclear reactors, Cr from chrome-plating dyes, Co from glass and ceramics, Cu from electrical goods, Mn from fertilizers, pharmaceutical, Hg from chromalkali industry etc. The concentration and distribution pattern of metals within the ecosystem is reflected by the biotic communities living in the system. WHO has laid down the following limits (values in mg/l) for heavy metals in drinking water : As

(0.05), Cd (0.005), Cr (0.05), Pb (0.05), Hg (0.001) and Se (0.01). Crossing of above limits adversely affects the consumer's health (Table 1).

Table 1: Limits prescribed for heavy metal content in Drinking Water in India.

Metal/ Metalloid	Concentration (mg l^{-1})	
	Permissible	Excessive
Arsenic	-	0.2
Chromium	-	0.05
Copper	1.0	3.0
Iron	0.3	1.0
Lead	-	0.1
Manganese	0.1	0.5
Selenium	-	0.05
Zinc	5.0	15.0

Metals are non-biodegradable hence once released into the environment, become an integral part of the habitat. They persist over a long period of time in the sediments of rivers and lakes by associating with organic and inorganic matters through the process of adsorption, complexation, chemical coordination. Metals also occur in a variety of forms as ions, compounds and complexes in the aquatics. Further, since metals can be accumulated by aquatic organisms, they can enter the food chain leading to man and thereby become a human health problem. For these reasons heavy metals continue to be of primary concern to environmental biologists. According to U.S. Environmental Protection Agency (EPA) prevention of pollution can be achieved by minimizing the use of materials, processes or practices and by eliminating the generation of pollutants or wastes at the source.

The best approach to restoration could be the halting of industrial proliferation or elimination of industries. But industries are coexistent or synonyms of development of society both in the developed and developing countries hence cannot be stopped. Therefore, large number of physico-chemical, biological and biotechnological methods are employed for restoration of metal damaged freshwater ecosystem. Physico-chemical methods include chemical precipitation, solvent extraction, electro-deposition, ion-exchange, ultrafiltration and activated carbon adsorption system for removal of heavy metals. All these methods are extremely expensive. Therefore, biological methods seem to hold considerable potential for use as alternative strategy.

The algae, bacteria, fungi, mosses, macrophytes, and other aquatic plants play a critical role in the decontamination of sewage and wastewaters and consequently in the functioning of aquatic environment. They can accumulate and concentrate heavy metals to levels substantially higher than those found in the surrounding environment. Due to the presence of different metal binding sites on the cell walls of cyanobacteria and bacteria, metals are adsorbed on their surfaces.

Biotransformation and environmental remediation techniques are used in modern biotechnology. Immobilized cell technology attracts considerable attention because of many advantages. Immobilized algal biotechnology is being increasingly used for removing metals from contaminated ecosystems. Metal binding proteins, phytochelatins, in plants are another detoxifying mechanisms. Keeping in view the increasing demand of clean water and continued shrinking of natural aquatic resources due to contamination, a need for restoration has been realized. Nevertheless, it merits mention that, the philosophy of restoration has been accepted by the Ministry of Science and Technology, Government of India with all seriousness and as a result it is now the part of thrust area programmes of Environmental Biotechnology.

2 Heavy Metals and Their Sources

Metals having specific gravity greater than 5 gm/cm^3 are generally known as heavy metals. Nieboer and Richardson [1] proposed an alternative classification of metals based on their coordination chemistry. Metals and metallioids are classified into class A (oxygen seeking), class-B (sulphur seeking) and borderline (intermediate between class A & B) elements. Class A contains alkali and alkaline earth elements (K^+, Na^+, Mg^{2+}, Ca^{2+}) etc. Class B comprise Cu^+, Hg^{2+}, Ag^{2+}, and Pb^{2+} which are very toxic. Borderline elements include Fe^{2+}, Fe^{3+} ,Mn^{2+}, and Cu^{2+}. Large number of heavy metals are required in trace concentrations by biological organisms, algae included, they however, become toxic if taken in high concentrations.

It is *sine-qua-none* that high concetrations of all heavy metals are toxic to algae, the extent of toxicity differs for different heavy metals[2,3,4]. Metal induced inhibition of algal growth follows the following trend : Hg>Cu>Pb>As>Zn = Cd>Ni, Cr, Sb, As. Photosynthesis is also affected by metals. Metals interfere with the regulation of DNA synthesis by blocking the -SH group[4].

Input of heavy metals into aquatic environment occurs by two sources (i) geologic weathering process, (ii) anthropogenic process. Geothermal discharges especially if they are saline or at low pH or high temperature can contain considerable amount of heavy metals. Fairly high concentration of different metals have been reported to occur in marine and fresh water ecosystems (see Table 2). Various sources of heavy metal input into inland waters include the following.

2.1 Aerial contamination :

The metal contaminated dust blown off the land remains in the atmosphere and contaminates the water bodies after its fall out or settling in the aquatic ecosystem.

2.2 Industrial processing of ores and metals

During the processing of ores, metal bearing dust particles are formed which may be only partially filtered out by air purification systems. Water percolates through slag heaps and mine tailing dumped from the mining operation and enters lakes carring very high concentration of metals and thereby increasing the metal burden of receiving water body.

2.3 Acidification

Atmospheric acid deposition is responsible for an additional input of hydrogen ions, sulfate and nitrate (H_2SO_4 and HNO_3) to aquatic and terrestrial

Table 2: Occurrence of some heavy metals in sea and fresh waters.

Metal/ Metalloid	Concentration ($mg\ m^{-3}$)	
	Sea water	Fresh water
Aluminium	2.0	300.0
Antimony	0.2	0.2
Berillium	0.006	0.3
Cadmium	0.1	0.1
Chromium	0.3	1.0
Cobalt	0.02	0.2
Copper	0.3	3.0
Lead	0.03	3.0
Manganese	0.2	8.0
Mercury	0.03	0.1
Molybdenum	10.0	0.5
Nickel	0.6	0.5
Silver	0.04	0.3
Tin	0.004	0.009
Uranium	3.0	0.4
Vanadium	2.5	0.5
Zinc	5.0	15.0

Source : Adopted from Bower, H.J.M. : *Environmental Chemistry of the Elements,* Academic Press, London, (1979) .

ecosystems. Gaseous pollutants (SO_2, NO_2) which come down as acid rain lower the pH of water bodies, and bring about release of heavy metals bound to different materials in the aquatic ecosystem. Effluents from acid mines and other industries, on discharge, add to the burgeoning problem of acidity and metal toxicity.

2.4 Soil erosion

As a consequence of the loss of vegetational cover the soil is subjected to severe leaching and erosion. Besides the increased acidity of soil increases heavy metal solubility and the runoff of such soils to aquatic milieu increases the metal load of such water.

2.5 Use of metals and metal compounds

Large number of metals and their compounds are used in different industrial processes. These include the use of chromium salts in tanneries, in compounds as plant protection agents, and tetramethyl lead as anti-knock agent. All these on discharge contaminates the aquatic ecosystem.

2.6 Leaching of metals from garbage and solid waste dumps

Mine dumps especially can be a serious source of pollution. Lead enters into aquatic environment from plating, mining, smelting, storage battery, printing, tetraethyl lead manufacturing, paint and dyeing industries and glass industrial operation[5]. Industries involved in mining and metal processing, dyeing and textile and petroleum refining cause Fe(II) pollution. Generally speaking, Cd is a non-essential element which enters into aquatic environment from mines, smelters and industries involved in the manufacture of alloys, paints, batteries and from burning of fossil fuel[6]. Volcanic action is also a major source of Cd. As, Hg in streams resulting from weathering of rock and soil. Mercury from flyash and/or fuel gas is continuously added to streams and lakes directly from the atmosphere and from washoff from the terrestrial environment. Microorganisms in the sediments convert the inorganic mercuricals to the more toxic dimethyl and methyl forms which are released into water, transported downstream or enter the aquatic food chain. Chromium is unique among the heavy metals found in industrial waste water and sewage sludge. It may exist as trivalent cation and as an anion in the hexavalent state in the pH range of agricultural soils.

3 Biological methods for Heavy Metal Removal

3.1 Algae and Cyanobacteria

Algae are regarded as the most suitable system for toxicity bioassay because they possess high accumulation potential due to their vast surface exchange area exposed to the environment. These organisms are widely used as biosorbent materials for the detoxification of metals because they are grown easily and cheaply and contain various metal binding functional groups. Cell wall composition is of importance in determining the interaction with toxicants. Some algal walls contain sulfated and acid mucopolysaccharides, others contain mucilages of various composition while the diatoms have silica associated with a pectin like material.

Accumulation of heavy metals to a level greater than that in the external environment appears widespread in aquatic organisms. Two mechanisms for controlling the concentration of toxic metals by aquatic organisms are ion exchange and complex formation. Heavy metal uptake is very complex and dependent on the metal ions and the biological system. In algae, cyanobacteria and aquatic plants metal uptake is biphasic i.e. initial rapid uptake (passive uptake) followed by a much slower energy dependent uptake (active uptake). During passive uptake, the metal ions are adsorbed on the surface of the cells within few seconds or minutes. This includes the physical adsorption, ion exchange and chemosorption. In the second phase, metal ions are transported across the cell membrane into the cytoplasm. Both passive and active transport processes have been proposed to account for heavy metal accumulation in algae.

A vast literature has been accumulated on the sequestering and accumulation of metallic ions by algae. Studies by Myers *et al*[7]. On phytoplankton demonstrated that the outer surface of the cells are negatively charged and Davies[8] suggested that the cellular surface of phytoplankton consisted of a mosaic of cationic and anionic exchange sites. The net charge on the cell surface is related to pH and other properties of the surrounding medium. Biosorption is an alternative method to conventional metal recovery. The metal binding capacity of copious sheath material of blue-green alga *Chroococcus paris* for removal of Cu, Cd and Zn, is well known[9]. Baxter and Jensen[10] reported uptake and accumulation of Mg, Sr, Ba and Mn by *Plectonema boryanum* in polyphosphate bodies. The formation of membrane whorls in Cd-treated cells of *Plectonema boryanum* has also been taken as one of the metal detoxifying mechanisms as lipoprotein may accomodate excessive amounts of heavy metals[11]. Green microalga *Stichococcus bacillaris* has the capacity to adsorb Cd[12]. Les and Walker[9] found that the metal binding increased significantly with a pH rise from 4-7. It has been found that heat killed *Chlorella* cells accumulated more Cd than live cells[13]. Thus they concluded that Cd uptake was not directly mediated by metabolic processes but by physico-chemical adsorption on the cell surface. By comparing the live and heat killed cells of *Anabaena cylindrica, Anabaena flos-aquae* and *Nostoc* spp. it was suggested that heat killed cells depicted highest metal (Ni) binding capacity than live cells[14].

Maeda *et al.*[15] studied bioaccumulation of Zn and Cd in the freshwater alga *Chlorella vulgaris.* The bioaccumulation and methylation of arsenic by freshwater algae are also applicable to the removal and detoxification of environmental arsenic. Andreas[16] described biosorption of five metals (Cd, Cu, Ni, Pb, Zn) by two different marine biomass i.e. *Sargassum fluitans* and *Ascophyllum nodosum.* Aksu *et al.*[17] observed that dead cells of *Chlorella vulgaris* and *Zoogloea ramigera* were good biosorbents for Cu and both pH and temperature were important in determining how well Cu could be bound to these species. Several other metal binding sites have been suggested for algae. Silverberg[18] found that dense intramitochondrial granules were formed in three green algae upon exposure to Cd. Mclean and Williamson[19] found that Cd was accumulated in the nucleus of the red alga *Porphyra.* For the alga *Microcystis aeruginosa,* Allen *et al.*[20] showed that, the toxicity of Zn^{+2} and $Zn(OH)^{+}$ was higher in comparison to chelated species.

Certain seaweeds, particularly the phaeophyceae and the Rhodophyceae are of significant commercial importance due to polysaccharide contents such as alginate and carrageenan. These have received special attention recently for their capability of accumulating metals from the aqueous solution. More recently, attention has turned to the usage of nonliving microorganisms as a potential industrial tool for metal removal and/or recovery. Greene *et al.*[21] revealed that algal biomass is a useful material for metal recovery. For metal ions including Cu(II), U(VI), Pb(II), Zn(II), Ni(II), Cr(II) and Cd(II), binding is believed to involve complexation of metal ions with ligands in or on the cell wall.

The biomass of *Ascophyllum nodosum* is one of the main sources of alginate/algin. The carboxyl groups of each polymer segment may play an important role as sites of cation binding. Co uptake by alginates was found to be proportional to the quantity of manuronic acid residues. The initial rate of cobalt biosorption gradually increased with increasing initial concentration of Co at pH 4 and at room temperature[22]. Twenty minutes were required for full biomass saturation at high initial concentration (600 mg l^{-1}) and excess of 10 min for low initial concentration (i.e. 100 mg l^{-1}). Within 5 min of contact, the biosorption system reached over 50-60% of the total Co taken up by the biomass.

The extracellular slime layer of cyanobacteria is known to interact strongly with cations. *Microcystis* slime layers consist predominantly of galacturonic acid and smaller amounts of neutral sugars and protein. In most other cyanobacteria, however, there is at least one uronic acid and several neutral sugars often in combination with protein[23].

Nealson and coworkers[24] reported binding of Mn with capsules of *Microcysti* in lake Oneida, New York and Green Bay, Wisconsin, USA. They suggested that *Microcystis* colonies after binding and oxidation of Mn become very heavy. Since manganese oxides are very good at binding other metals viz., Cd, Ni, Zn, Pb etc. the formation of such manganese oxides by *Microcystis* may strip many toxic metals from the water. Following mechanism was proposed for Mn(II) oxidation by *Microcystis.*

1. Adsorption or accumulation of Mn (II) by capsule

Mn (II)
Capsule ---------------------- Capsule-Mn (II)

Capsule adsorbs soluble Mn, accumulating it in close proximity to cells.

2. Products of photosynthesis accumulate within capsule

Cells ------------------------- Oxygen accumulation, pH alteration

3. Oxidation of Mn (II)

high pH, high Eh
Capsule-Mn (II) -------------------- Capsule-MnOx

High pH conditions lead to conversion of Mn(II) to Mn(IV), which is a strong acid and forms insoluble oxides in presence of oxygen.

Rai and his associates are currently working on the metal binding capacity of capsulated and capsule depleted *Microcystis.* Capsule depleted *Microcystis* accumulated only 60-70 % of Cu, Ni and Cd as compared to capsulated *Microcystis* (See Figs. 1 and 2). By Comparing wet biomass with that of air

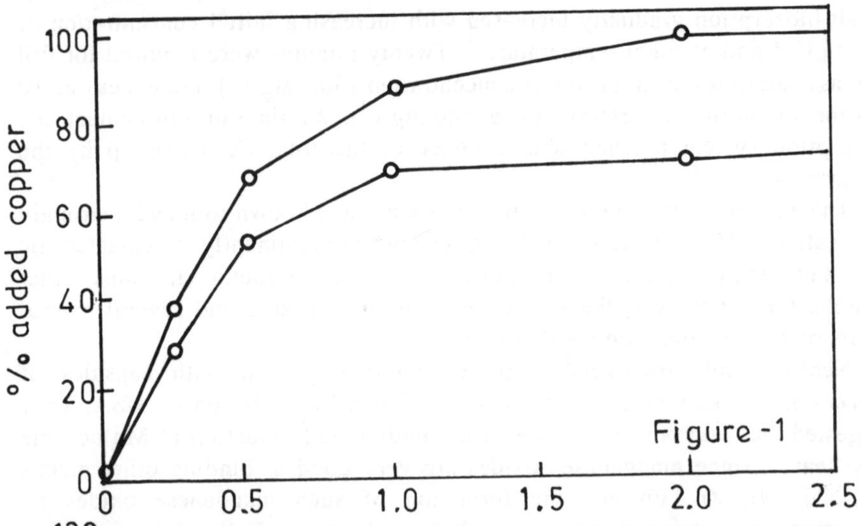

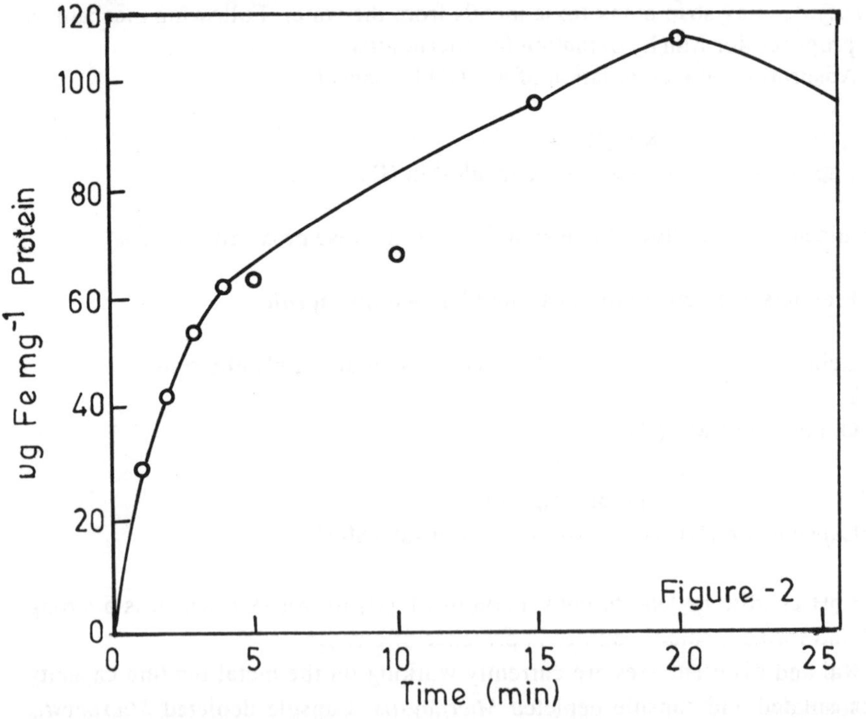

Figure 1 : Uptake of copper by capsulated (-) and capsule depleted (O-O) *Microcystis*.

Figure 2 : Adsorption maxima of iron in capsulated *Microcystis*

dried biomass, it was found that air dried biomass depicted a significant reduction in metal accumulation i.e. 35, 28 and 25% respectively for Cu, Ni, and Cd (See table 3b).

Table 3(a) : Effect of cell viability on Copper accumulation

Treatment	Accumulated copper	
	$\mu g\ ml^{-1}$	% of Control
29^0 C : 2100 lux	0.209±0.011	100
29^0 C : Dark	0.198±0.009	95
04^0 C : Dark	0.181±0.013	87
Heat killed	0.181±0.011	87
0.1% HCHO	0.204±0.022	98

Table 3 (b) : Metal accumulation in ug metal per mg biomass (dry weight) per hour

Material	$\mu g\ Ni\ mg^{-1}\ hr^{-1}$	$\mu g\ Cd\ mg^{-1}\ hr^{-1}$	$\mu g\ Cu\ mg^{-1} hr^{-1}$
Wet biomass	0.103±0.011	0.075±0.007	0.080±0.007
Dry biomass	0.075±0.009 (28)	0.054±0.001 (25)	0.052±0.005 (35)

All the values are mean ± SE.
Data in paranthese denote % reduction.

The heat killed and the dark incubated cells at 4°C showed only 13% lower Cu accumulation as compared to the control (See Table 3a). Their study has therefore demonstrated that capsule plays significant role in metal adsorption.

3.2 Aquatic plants

All plants have the ability to absorb, from soil and water, those heavy metals which are essential for their growth and development. These metals include Fe, Mn, Zn, Cu, Mg, Mo and Ni. However, many plants have the ability to accumulate such heavy metals as Cd, Cr, Pb, Co, Ag, Se and Hg [25,26], which have no biological function. The basic idea that plants can be used for environmental remediation is certainly very old and cannot be traced to any particular reference. For example, extensive research on using semi-aquatic plants for treating radionuclide contaminated waters existed in Russia during the dawn of the nuclear era[26]. The knowledge that aquatic or semiaquatic vascular plants such as water hyacinth i.e. *Eichornia crassipes*[27], Pennyworth i.e. *Hydrocotyle umbellate*[28], duckweed i.e. *Lemna minor*[29,30], and water velvet i.e. *Azolla pinnate*[29], can take up Pb, Cu, Cd, Fe and Hg from contaminated solutions did exist for a long time. This ability is currently utilized in many constructed wetlands, which may be effective in removing some heavy metals as well as organics from water[31].

3.3 Macrophytes

There are number of promising biosorbent candidates for removing targeted metals and radionuclides from dilute solution. Macrophytes, besides algae, are the main primary producers in ponds and lakes. These are usually found in metal-enriched wastewaters and tend to concentrate metals to exceptionally high level. A direct relationship between metal content in the milieu and organisms has been reported and suggestions have been put forth to use metal content of plants for biomonitoring of metals in aquatic environments.

Free-floating *Lemna polyrrhiza* and widely distributed water fern *Azolla pinnata* capable of fast growth and easy to handle may be employed for stripping nutrients from wastewater. Roots usually accumulate significantly higher metal levels than the above-ground plants[32] (See Table 4).

High quantities of Cd, Cu, and U were localised within the cell wall of the shoot and root of *Azolla*[33]. In *Eichornia crassipes* and *Pistia straliotes* the concetration of As in the roots was higher than that in the leaves[34]. All these reports tend to indicate that roots can immobilize heavy metals and provide very efficient barrier against heavy metal translocation within the plant.

Table 4: Analysis of Water Hyacinth before (X) and after (Y) exposure in effluents containing metals

Metal	Exposed for six weeks (mg l^{-1})					
	Leaves		Stems		Roots	
	X	Y	X	Y	X	Y
Cadium	0.1	2.0	0.1	10.0	0.1	165
Copper	16.5	32.0	10.1	45.0	24.0	595
Lead	8.4	35.0	2.1	44.0	40.0	290
Silver	0.8	9.0	0.1	5.0	36.0	110
Nickel	0.1	1.0	0.1	9.0	0.1	160
Chromium	0.1	4.0	0.1	12.0	0.1	280

Source : Adopted from Bower, H.J.M. : *Environmental Chemistry of the Elements,* Academic Press, London, (1979) .

Enhanced efficiency of dead plants in adsorbing high levels of metals could be explained in terms of increased availability of binding sites in the intracellular compartment due to the break down of permeability barrier. Dead biomass of *Azolla filiculoides* has a very high capacity to bind heavy metals[35]. The efficiency of dead biomass of *Azolla pinnata* and *Lemna polyrrhiza* in adsorbing Cd at different time interval was high as compared to live biomass. Utilization of these plants for alleviating heavy metal burden from wastewater is well known.

3.4 Bryophytes and Lichens

Extracellular cation uptake by bryophytes and lichens is an ion exchange phenomenon involving reversible binding to anionic sites outside the protoplast[36]. The capacity of some lichen sps. to accumulate high concentration of toxic elements has led to their use as pollution monitors[37]. Lichens accumulate metals from airborne particle or from materials dissovled and suspended in the aqueous solution that come into contact with them.

Intracellular and extracellular uptake of Cd was investigated into species of the lichen genus *Peltigera*[38]. Some variation in the capacity of *Umbilicaria* spp. to take up Ni has been demonstrated by Richardson and Nieboer[39] who also showed that *Peltigera* species have more than three fold higher Ni uptake capacity than that of *Umbilicaria* spp. Approximately 75% of 100 mol m^{-3} Ni taken up by *Umbilicaria muhlenbergii* occured witin 5 min of exposure. The multilayered structure of the lichen increased the size of boundary layer through which metal ions must diffuse before being bound to the surface.

Studies on aquatic mosses and liverworts from waters contaminated with heavy metals indicate their ability to accumulate such metals to high levels[40]. Input of heavy metals to rivers leads to increase in the levels of these metals in all the sps of bryophyte. *Fontinallis squamosa* and *Rhynchostegium riparioides* are useful monitors of heavy metal pollution in freshwater. *Scaparia undulata* has considerable attraction as a monitor of environmental levels of heavy metals. It can tolerate very high levels of Zn, usually together with elevated Cd concentration.

3.5 Bacteria

The study of bacterial sorption of metal ions is receiving increasing attention mainly due to its environmental and technological implications. The high metal uptake capacity of many bacteria has encouraged the application of microbial biomass for detoxification of effluents as well as for metal recovery.

The cell wall of *Bacillus subtills* and many other Gram-positive bacteria consists of a highly organized collection[41,42] of anionic hetero and homopolymers which are mostly polysaccharides. On the basis of the chemical composition and cell organisation, bacateria and blue-green algae are very similar. A peptidoglycan polysaccaride complex of the cell wall of filamentous prochlorophyte *Prochlorothrix hollandica* is similar in chemical composition and structure to that found in cyanobacteria.

The metal binding in bacteria appears to be at least a two-step process, in which the first event is a stoichiometric interaction between metal and reactive chemical groups in the wall fabric. The next event is an inorganic deposition of increased amounts of metal, which can be readily visualized by electron microscopy due to the electron scattering by the walls.

Little is known about the sites of metal deposition in the wall, though the phosphodiester groups of teichoic acid have been implicated in the binding of

Mg[43]. This study attempts to locate the active sites by comparing the metal uptake values of native *Bitolic subtilis* walls with those in which the electrochemical charges (reactive groups) of various wall constituents have been chemically modified or removed. In addition, the nature of the metal deposit has been studied by X-ray diffraction and electron scattering ability of the cell by the following reactions[44].

$$\underset{\text{wall}}{* \text{-}NH_3^+} + \underset{\text{iodoacetate}}{ICH_2COO^-} \text{ ------------------- } * \text{-}NH_2\text{-}CH_2COO^- + HI$$

Reaction. The reaction between the bacterial cell wall and soidum iodoacetate.

N_2^- fixing bacterium *Azotobacter vinelandii* contains abundant carboxylic group which could provide several binding sites for metal ions. *A. vinelandii* is capable of immobilizing metal ions from aqueous solution. This biosorption is a reversible process affected by the pH of the solution and the presence of competitive cations[45]. An increased removal of Cu^{2+}, Cd^{2+}, and Zn^{+2} at high pH has been observed using polygalacturonic acid, a natural anionic polysaccharide obtained from pectin[46]. From the technological view point this strain shows a lower binding capacity than that observed for other microorganisms, but its relative high affinity could be useful in a secondary treatment of mining or industrial wastewaters.

Sulphate reducing bacteria constitute a special gruop of microorganisms that use sulphate as terminal electron acceptor for their respiration. Their ability to generate large quantities fo H_2S is important for a variety of industrial applications. Panchanandikar and Kar[47] studied the precipitation of Cu using *Desulfovibrio* sps. Hg is one of the metals given high priority for removal from wastewaters because of its high toxicity. Recently biological processes for the removal of Hg from waste waters have been investigated. A process based on bioaccumulation of Hg[48] by genetically modified, mercury resistant *Pseudomonas putida, Aeromonas hydrophila* and natural consortia has been developed in a bench scale column. Continuous culture of Hg-resistant *Pseudomonas putida* strain FB-1 can convert inorganic Hg(II) to gaseous elemental Hg(o) enzymatically[49] and remove from the environment. This bioprocess appears better than the bioaccumulation process since waste is processed and the cost of sludge dumping which is becoming economically prohibitive in waste treatment could be lowered.

3.6 Fungi and Yeast

The uptake of heavy metals is of fundamental importance to organisms growing in polluted habitats since tolerance may be deternined by the ability to prevent or reduce cellular entry and/or the ability to compartmentalise or detoxify it within the cell. It is apparent that the cell walls of yeast and chlamydospore of polymeric fungus *Aureobasidium pullulans* play important roles both in the uptake and prevention of heavy metal toxicity. Protoplast prepared from yeast like cells, hyphae and chlamydospores of *Aureobasidium pullulans* can take up heavy metals as Zn^{2+}, Co^{2+}, Cd^{2+}, and Cu^{+2}. The ability of the thick-walled, melanised chlamydospores to bind much larger amount of heavy metals than hyaline cell type[50], and the use of protoplasts has confirmed that the chlamydospore cell wall does act as a barrier towards metal ions, preventing entry into the cell. For *Aureobasidium pullulans,* hyaline hyphae and yeast like cells are capable of intracellular uptake of Cu, Co, Cd, Zn whereas chlamydospores only exhibit binding to cell walls and associated material and are more metal-tolerant than the former cell types[51].

Studies with fungal and yeast biomass have shown effective uptake for a range of metal ions including copper, cobalt, caesium, strontium and uranium with variety of mechanisms being implicated in thier uptake[52,53]. Louise and Gadd[54] studied biosorption of uranium, strontium and caesium by pelleted mycelium of *Rhizopus arrhizus* and *Penicillium chrysogenum.* The immobilized *Saccharomyces cervisiae* was evaluated in both batch and continuous flow systems where presence of competing cations affected accumulation. For metal removal and recovery dead fungal biomass seems to offer several advantages over live biomass: (I) it may be obtained cheaply from several industrial sources, (ii) needs no nutrient supply, and (iii) recovery of surface bound metals may be relatively a simple non-destuctive treatment[55,56].

Cross flow microfiltration was shown to retain *Saccharomyces cervisiae* utilized for heavy metal accumulation. The passage of metal-laden effluent through a series of sequential bioaccumulation systems allowed for further reduction in the levels of Cu, Cd and Co in the final effluent than that afforded by a single bioaccumulation process[57]. Serial bioaccumulation systems also allowed for partial separation of metals from dual metal influents. More than one elemental metal cations could be accumulated simultaneously and in greater quantity than when a single metal was present in the effluent.

4 Biotechnological Methods

Toxic metal contamination of aqueous waste water is still in need of an effective and affordable technological solution. The bioremediation paradigm, "using microorganisms to degrade pollutants in situ" has recently been attracting a lot of public attention. Unfortunately, heavy metals and radionuclides (differ in the mass of metals) cannot be chemically degraded. Therefore application of microbial bioremediation to the in situ removal of heavy metals from contaminated substrates is mainly limited to their immobilization by precipitation or reduction.

Several techniques were developed that allowed the *in vitro* fragmentation and rejoining of molecules of deoxyribonucleic acid (DNA) and the placement of these molecules into living cells[58]. Use of these recombinant DNA techniques led to the development of biotechnology, broadly defined as the application of biological organisms and their products to commercial process (Office of Technology Assessment, OTA, 1988). Bacteria have been used more than other organisms because of their short generation time, lack of a defined nucleus, and relatively simple genetics as compared to eukaryotes. Genetic engineering is used to develop organisms capable of performing commercially desired tasks, including producing pesticides and pharmaceuticals, fixing nitrogen, leaching ores, recovering oil, degrading hazardous wastes and constructing biological control agents for pests or diseases. Biotechnology products will, one hopes, benefit human society.

4.1 Immobilized technology in metal removal

Although free cells do accummulate metals, the immobilized cell technology is receiving greater attention of environmental biotechnologists. Cell immobilization is the one possible method for enzyme immobiliation. It can be used to overcome the problems and limitations (co-factor regeneration and arrangement of enzyme molecules in ordered clusters) frequently encoutered in direct enzyme immobilization. According to Abbott[59] immobilization is considered as a physical confinement or localization of microorganisms that permits their economic reuse. Entrapment within gel forming polysaccharides appears to be the most suitable method for immobilization.

Microorganism and other cellular materials have a natural inclination to adhere to surfaces and in this way, they become immobilized. Metal accumulation by microbial and algal cells shows number of disadvantages due to their small particle size, low mechnaical strength and low density which limit the choice of reactor systems and make biomass or effluent separation difficult[60]. In comparison to free cells, immobilized cells offer certain specific advantages :

(a) Renewable or self proliferating biocatalysts protected against unfavourable environmental factors outside their functional area.
(b) Decreased cell metabolism and cell wall permeability.
(c) Immobilized cells can be used for repeated cycles.
(d) The physcio-chemical interactions which take place between carriers and cells often give rise to increased stability and cell productivity.
(e) The density per unit volume of immobilized cells is higher than that of freely suspended cells, so that faster reaction rate is achieved by using the immobilized cells.
(f) Reduction of cost due to easy separation of cells and excreted products.
(g) No washout of cells.

The choice of matrices and immobilization procedures will depend on the nature of the substrates used and products formed and also on the reaction conditions. The materials commonly used for immobilization are sodium alginate, agarose, chitosan[61,62,63], synthetic fibres, glass beads, polyvinyl foam, polyurethrane foam[64], silica gel, nylon, photo-cross linked resin, albumin, and agar, pectate[65,66].

Chevalier and de la Noue[67,68] have shown that cells of *Scenedesmus* immobilized on K-carrageenan beads are as efficient as free cells in removing NH_4^+ and PO_4^{3-} from secondary urban effluents. On comparing the performance of chitosan immobilized and free suspension of *Phormidium* in tertiary treatment, Pouliot *et al.*[69] found that immobilized algae posessed a high capacity to deplete inorganic nitrogen, and orthophosphate. Such algae are not only resistant to high dosage of toxic metals but also possess high chlorophyll content[70] and greater efficiency than free cells in removing metal and nutrients over repeated cycles[62]. Rai and Mallick[62] demonstrated that after using, *A. doliolum, Chlorella vulgaris* once, when taken out, washed and reinoculated in fresh medium containing same concentration of Cu and Fe, the metal removal efficiency was decreased by 12 and 19 percent from first to second and by 29 and 36 percent from second to third cycle. Rai and associates further demonstrated that pH and presence of cations significantly affect the metal removal efficiency of immobilized algae. It has been critically demonstrated that a pH between 6-8 or neutral, beads with low cell density, presence of humic acid (having various metal binding sites) are some of the factors favouring the removal of metals[63]. Attempts have also been made to find out a cost effective matrix for immobilization of algal cells. By comparing the removal efficiency of algae immobilized in alginate, carragennan, chitosan and agar-agar, maximum efficiency for scavanging Cr, Cu, Zn and Ni was observed with chitosan[71].

4.2 Phytochelatin

To resist the toxic effects of heavy metals, the plants must either limit their cellular uptake[72], detoxify the heavy metals that enter the cells or develop heavy metal resistant metabolisms. Some plant ecotypes endemic to heavy metal polluted soils have been shown to contain metal resistant enzymes e.g. cell wall acid phosphatase[73]. However, it is unlikely that the development of heavy metal resistant biochemical processes could be a viable metal resistance mechanism. Once accumulated within cell, metals will need to be detoxified. This can occur in a number of ways, depending on the metal, either through chelation, compartmentalization or precipitation. For instance, Zn may be chelated to organic acids and accumulated within the vacuole[74]. Intact vacuoles isolated from tobacco and barley exposed to Zn were shown to contain high levels of Zn[75] . Zn accumulation within the vacuole, as a Zn detoxification mechanism, is supported by the observation that the vacuolar volume fraction of meristematic cells within the root tip of *Festuca rubra* increases during Zn exposure[76]. Another mechanism for the detoxification of intracellular Zn is its precipitation as Zn-phtpate[76,77,78,79].

Cd is also known to accumulate within the vacuoles[79] where it associates with the family of thiol-rich peptides, called phytochelatins[80,81]. The recent discovery of mechanisms for the transport of both Cd and Cd-phytochelation complex across the tonoplast[82,83] supports the suggestion that Cd detoxification is achieved by the accumulation of Cd within the vacuole. Experiments have demonstrated that in roots of *Brassica juncea* majority of Cd is chelated with sulfur, the major chelating group in phytochelatins. The Cd phytochelatin complex also contains inorganic sulfide[84,85] which is thought to stabilize and increase its Cd binding capacity.

4.3 Phytoremediation

The use of specially selected and engineered metal accumulating plants for environmental clean-up is an emerging technology called "Phytoremediation". It is a new field, and has great potential. This will require a multidisciplinary approach, spanning over fields as diverse as plant biology, agricultural engineering, agronomy, soil science, microbiolgy and genetic engineering. Three subsets of this technology are applicable to toxic metal remediation.

4.3.1 Phytoextraction

In this case the metal accumulating plants are used to remove toxic metals from soil. The plants transport and concentrate metals from the soil into the harvestable structures such as roots and aboveground shoots[86]. The plants used for phytoextraction have a rapid growth rate and the potential to produce a high biomass in the field.

4.3.2 Rhizofiltration

The use of metal accumulating plants to remove toxic metals from polluted waters in which plant roots absorb, precipitate and concetrate toxic metals from polluted effluents[87]. Mechanisms of toxic metal removal by plant roots are different for different metals. In the case of Pb, sorption by the root is probably the fastest component of metal removal. Rhizofiltration is particularlyl effective and economically spell binding when low concentration of contaminants and large volume of water are involved.

4.3.3 Phytostabilization

The heavy metal tolerant plants are used to eliminate the bioavailable toxic metals from soils, thereby reducing the risk of further environmental degradation by leaching into the ground water or by air borne spread[88]. For the stabilization of metaliferous wastes in UK, Smith and Bradshw[89], utilized local metaltolerant plant species. Plants may also reduce metal leaching by converting metals from a soluble oxidation state to an insoluble oxidation state. A good phytostabilizing plant should tolerate high levels of heavy metals and immobilize these metals in the soil via root uptake, precipitation or reduction.

Phytoremediation of heavy metals is designed to concentrate metals in plant tissues, thus minimizing the amout of solid or liquid hazardous waste which needs to be treated and deposited at hazardous waste sites, with the ultimate goal of developing an economical method of reclaiming metals from plant residue. This will completely eliminate the need for costly off-site disposal.

4.3.4 Future Perspectives

Use of genetic engineering techniques can help in improving the metal removal efficiency of microbes and plants. For instance, genes encoding the C-binding protein, metallothionein, have been expressed in plants, is a seemingly successful attempt to increase Cd-resistance[90]. Another strategy for improving the phytoremediation potential of high biomass plant species is the introduction of genes responsible for metal accumulation and resistant from the wild metal accumulators. In the absence of known phytoremediation genes this may be accomplished in a somatic and sexual hybridization followed by extensive screening and backcrossing of progeny. However, a long term effort should be directed towards developing a “molecular tool box” which is composed of genes responsible for phytoremediation. Systematic screening of plant species and genotypes for metal accumulation and resistance will broaden the spectra of genetic material available for optimization and transfer, Mutagenesis of selected high biomass plant species may also produce improved phytoremediating cultivars.

4.4 Engineered reed-bed systems for waste treatment

Over the past decade, constructed wetlands are used for removal of heavy metals[91,92,93]. The main types of aquatic plants used in constructed wetlands are the common reed (*Phragmites australis)*, the common reedmace (*Typha latifolia)* and the common club-rush (*Schoenoplectus lacustris)*. In constructed wetlands reeds are planted in soil, gravel matrix or pea-gravel etc. according to the types of bed, such as horizontal-flow reed-beds consist of soil or gravel matrix. Down-flow reed-beds are constructed of sharp sand, pea-gravel, coarser gravel and stones in layers of appropriate thickness. Above ground flow between plant stems is used for the removal of metals, and to bring about a pH change; this is particularly useful for the treatment of acid-mine discharge. Work carried out by Seidel[94] showed that significant removal of pollutants was possible when contaminated water was passed through beds of reed planted in soil or gravel.

The plants used in this technology have a root system of rhizomes, containing thick hollow air passages, from which fine hair roots hang down; the vertical aerial shoots develop upwards from the rhizome. Oxygen from leaves passes through the stems and rhizomes and exuded from the fine roots so that a thin, oxygenated, aqueous film less than 1 mm thick surrounds the hairs. Consequently, this root zone, or rhizosphere, can support a very large population of aerobic microorganisms, which are dominated by bacteria, have relatively low enzymic activities within the organic substrates tested[93], and would, therefore, be expected to degrade most of the simpler organic materials, i.e. those contributing most to the biochemical oxygen demand (BOD). By contrast, the fungi and actinomycetes, although fewer in number, have a wider range of hydrolytic activities[93].

Wastewater passing through the thickly developed reed-bed rhizosphere encounters alternate aerobic and anaerobic microbial populations. Microorganisms can form biofilms around the lower stems, which can then trap particles suspended in the waste water by adsorption.

4.5 Percentage removal of pollutants by reed beds

Bed type	Removal type
Horizontal subsurface flow	Consists of large number of beds, popular in Europe 80-90 % BOD removal with a typical outlet conc. of 20 mg^{-1}, total-N removal 20-30 % and total phosphate removal is 30-40%.[92,93].
Downflow	Using a system of two downflow stages to treat domestic Sewage achieved removal of 93 % BOD, 90 percent suspended solids (SS), 75 % NH_4-N and 37% orthophosphate[92].

Overland flow — On passing the effluent of a lead-zinc mine through a bed of *Typha latifolia,* the reduction of SS 99%, Pb 95% and Zn 80% can be achieved.[92]

The technology of constructed wetlands is still relatively new. The use of wetlands for the removal of metals has great potential. It will be needed for the treatment of seepage from abandoned coal and metal mines in different countries.

5 Microbial Biotechnology

5.1 Synthetic oligonucleotide probes for detection of mercury-resistance genes

Microbial communities adapt to toxicants through development of resistance and ability to degrade or utilize these substances. An important organo-mercurial contamination occurred in the Rhine-river following the extinction of a fire in a Swiss factory on November 1, 1986 during which various pesticides, fungicides and herbicides including organomercurials were washed into the Rhine river at concentration of total mercury between 5 to 10 ug/litre. Organomercurial compounds are well known for their high toxicity. Their detoxification by resistant bacteria involves the reduction of the organomercurial form (R-Hg^{+}) to the less toxic element mercury (Hg^{0})[95]. The organomercurial lyase enzyme encoded by the mer B gen is involved in the cleavage of R-Hg^{+} bond to release Hg^{2+}. Mercuric reductase, the enzyme encoded by the mer A gene, is involved in the conversion of Hg^{2+} to Hg^{0}.[96] Protozoa, and especially Amoeba, have received little attention in pollution studies although they play an important ecological role[97].

Available nucleic acid technology allows the detection of specific DNA sequences in the genomes of various microorganisms. This approach can be utilized for the detection of genes coding for specific enzymes responsible for pollutant biodegradation. The development of DNA gene probes for detection of Hg-resistance genes in bacteria was achieved by using constructed DNA restriction fragments from plasmid R100. However, the successful development of gene probe requires that any non-specific DNA sequences be separated from "mer" sequence, to prevent background hybridization. Mirgain *et al.*[98] used published "mer A" and "mer B" gene sequences to synthesize specific oligonucleotide probes. The "mer" genes are well conserved amongst bacterial strains and are commonly carried on plasmids or tranposons which are favourable for the genetic material within bacterial comunities[98]. Synthesizing short oligonucleotide sequences, chosen for their absence of homologies with other enzymes, secured an accurate detection of "mer A" and "mer B" genes.

5.2 Genetically engineered Pseudomonas : Application in Bioremediation

A genetically engineered microorganisms (GEM), *Pseudomonas* sp. B13 FRI (pFRC 20P) (abbreviated FRI20), has previously been developed to simultaneously mineralize mixtures of methylated and chlorinated benzoic acids and phenols through a modified orthocleavage pathway. Its performance was investigated both in different types of aquatic microcosm and in pure culture to determine the following.

(i) If under simulated conditions the genetically engineered pathway effectively removes mixtures of metal pollutants simultaneously, quickly and completely,

(ii) Where the optimum pollutant concentration range for this activity lies,

(iii) How physical, chemical and biological factors in the microcosms influence degradation rates?

Growth and degeradation parameters of FRI20 in pure culture were determined with 3-chlorobenzoate (3CB), 4-methyl benzoate (4MB) and equimolar mixtures of both as carbon sources. These substrates were degraded simultaneously, albeit with different degradation velocities, by FR120. Environmental conditions in the sediment were more favourable for the degradation of substituted aromatics than those in pure culture. The physiological characteristics of FR120 and its performance in aquatic microcosms makes it a good candidate for bioremediation at sites contaminated with mixtures of chlorinated and methylated aromatics.

Acknowledgement :

We are thankful to the Department of Biotechnology, Ministry of Science and Technology, New Delhi for financial assistance.

References

1 E. Nieboer, and D.H.S. Richardson, *Environ. Pollut.* **1**, 3 (1980).
2 L.C. Rai, J.P. Gaur, and H. D. Kumar, *Biol. Rev.* **56**, 99 (1981).
3 B.A. Whitton, in, *Algae as Ecological Indicators,* ed. L.E. Shubert, Academic Press, London (1984).
4 L.F. De Filiopis, and C.K. Pallaghy, in, *Algae and Water Pollution.* eds. L.C. Rai, J.P. Gaur, and C.J. Soedor, E. Schweizer, Verlag. Stuttgart, (1994).
5 K.P. Yadava, B.S. Tyagi, V.N. Singh, *J. Chem. Technol. Biotechnol.* **51**, 47 (1991).
6 S. W. Reeder, A. Demayo, M.C. Taylor, Inland Water Directorate, Water Quality Br., Canada (1979).
7 V.B. Myers, R.L.Iverson, and R.C. Harriss, *J. Exp. Mar. Biol. Ecol.* **17**, 59 (1975).
8 A.G. Davies, *Adv. Mar. Biol.* **15**, 381 (1978).
9 Les Albin and R.W. Walker, *Water Air Soil Pollut.* **23**, 129 (1984).
10 M. Baxter, and Jensen, *Protoplasma* **104**, 81 (1980).
11 J. W. Rachlin, T.E. Jensen, M. Bakter, and V. Jani, *Arch. Environ. Contam. Toxicol.* **11**, 323 (1982).
12 T. Skowroisk, *Chemosphere* **15**, 69 (1986).
13 T. Sakaguchi, T. Tsuji, A. Nakajima, and T. Horikoshi, *Eur. J. Appl. Microbiol. Biotechnol.* **8**, 207 (1979).
14 S.L. Corder, and R. Mark, *Appl. Biochem. Biotechnol.* **45/45**, 847 (1994).
15 Maeda, Shigeru., M. Makoto, O. Akira, and T. Toshio, *Chemosphere* **21,** 953 (1990).
16 L. Andreas, Z. R. Holan and B. Volesky, *J. Chem. Tech. Biotechnol.* **62**, 279 (1995).
17 Aksu Zumriye, Yesim Sag and Tulin Kutsal. *Environ. Technol.* **13,** 579 (1992).
18 B.A. Silverberg, P.M. Stokes, and L.V. Ferstenberg, *J. Cell Biol.* **69**, 210 (1976).

19 M.W. Maclean, and F.B. Williamson, *Physiol. Plant.* **41,** 268 (1977).
20 H. E. Allen, R. H. Hall and T. D. Brisbin, *Environ.Sci. Technol.* **14**, 441 (1980).
21 B. Greene, R. McPherson, and D. Darnall, in *Metal Speciation and Recovery,* eds. J.W. Patterson and R.Passino Chelsea, MI : Lewis (1986).
22 N. Kuyucak, and B. Volesky, *Biotechnol. Bioengg.* **33**, 823 (1989).
23 M. Nakagawa, Y. Takamura, and O. Yagi, *Agri. Biol. Chem.* **51**, 329 (1987).
24 K. Nieboer, B. Tebo, and R. Rosson, *Adv. Appl. Microbiol.* **33**, 279 (1988).
25 A. J. M. Baker and R. R. Brooks, *Ecol. Phytochem.* **1**, 81 (1989).
26 I. Raskin, P.B.A.N. Kumar, S. Dushenkov, and D.E. Salt, *Curr. Opinion Biotechnol.* **5**, 285 (1994).
26 E.A. Timofeev-Resovsky, B.M Agafonov, N.V. Timofeev-Resovsky, *Proceedings of the Biological Institute of the USSR Academy of Sciences,* **22**, 49 (1962).
27 F.E. Dierberg, T.A. De Busk, and N.A. Goulet, in, eds Reddy, K.B. and Smith, W.H. Mangolia Publishing Inc., Fl. (1987).
28 S.K. Jain, P. Vasudevan, and N.K. Jha, *Biological Wastes.* **28**, 115 (1989).
29 S.C. Mo, D.S. Choi, and J.W. Robinson, *J. Env. Sci. Health.* **A24**, 135 (1989).
30 P.J. Jackson, A.P. Torres, E. Delhaize, E. Pack, and S.L. Bolender, *J. Env. Quality.* **19**, 644 (1990).
31 T. Wildeman, and J.N. Cevaal, *The Hazardous Waste Consultant.* July/August: **1,** 24 (1994).
32 S. W. Breckle, in, *The Root System: The Hidden Half eds.* Y. Waisel, U.Katkaji, A. Eshel, Marcel Dekker Inc., New York (1989).
33 M. Sela, E. Tel-Or, E. Frizt, and A. Hutterman, *Plant Physiol.* **88**, 30 (1988).
34 C.K. Lee, K.S. Low, N.S. Hew, *Sci. Tot. Environ.* **103**, 215 (1991).
35 M. Sela, J. Gaity, E. Tel-Or, *New Phytol.* **112**, 7 (1989).
36 D.H. Brown, in, *The Experimental Biology of Bryophyles*, A.F.Dyer, and J.G. Duckett, Academic Press, London (1984).
37 M.H. Martin, and P.J. Coughfrey, *Biological Monitoring of Heavy Metal pollution.* Appl. Sci. Publi., London and New York (1982).
38 R. P. Beckett, and D. H. Brown, *New Phytol.* **97**, (1983).
39 D.H.S. Richardson, and E. Nieboer, *Lichenologis* **15**, 81 (1983).

40 P.J. Say, J.P.C. Harding, and B.A. Whitton, *Environ. Pollut. Ser.* **8**, 295 (1981).

41 H. Formanck, S. Formanck, and H. Wawra, *Environ. J. Biochem.* **46**, 279 (1974).

42 R.W.H. Verwer, and N. Nanninga, *Arch. Microbil.* **109**, 109 (1976).

43 T. J. Beveridge, F.M.R. Williams, and J. J. Koval, *Can. J. Microbiol.* **24**, 1439 (1978).

44 T. J. Beveridge, and R.G.E. Murry, *J. Bacteriol.* **141**, 876 (1980.

45 D. Cotoras, M. Miller, P. Viedma, J. Pimental, and A. Mestre, *World J. Microbiol. Biotechnol.* **8**, 319 (1992).

46 H.H.G. Jellinek, and S.P. Sangal, *Water Res.* **6**, 305 (1972).

47 V.V. Panchanadikar, and R.N. Kar, *World J. Microbiol. Biotechnol.* **9**, 280 (1993).

48 M. Brunke, D. Deckwer, A. Frishmutch, J. M. Horn, H. Lunsdorf, M. Rohircht, K.M. Timmis, and P. Weppen, in proceedings of the 19th International Symposium on Biohydrometallurgy eds J. Cardoso Duarte, and R.W. Lawrence, Forbitec Edition (1991).

49 F. Baldi, F. Parati, F. Semplici and V. Tandoi. *World J. Microbiol. Biotechnol.* **9**, 275 (1993).

50 J.L. Mowll and G.M.Gadd, *J. Gen. Microbiol.* **130**, 279 (1984).

51 G. M. Gadd, and J.L. Mowll, *Exp. Mycol.* **9**, 230 (1985).

52 G. M. Gadd, and C. White, in, *Metal Microbe Interactions eds.* R.K. Poole, and G.M. Gadd, IRL Press, Oxford (1989).

53 G.M. Roomans, A.P.R. Theuvenet, T.P.R. Vanden Berg, and G.W.F. Borst-Pauwels, *Biochim. Biophys. Act* **551**, 187 (1979).

54 Louise de Rome and G.M. Gadd, *J. Indusst. Microbiol.* 7, 97 (1991).

55 G. M. Gadd, in, *Microbes in Extreme Environments. eds.* R.A. Herbert, and G.A. Codd, Academic Press, London (1986).

56 G. M. Gadd, in, *Biotechnology* ed. H. J. Rehm, eds. vol 6b. VCH Verlags gasellschft. Weinheim. (1988).

57 D. Brady, R. D. Rose, and Dunan, *Biotechol Bioengg.* **44**, 1362 (1994).

58 J. Cairns Jr., and D.R. Drvos, *Reviews of Environ. Contamination and Toxicol.* **124**, 19 (1992).

59 B. J. Abbot, Annual reports on Fermentation Processes. **1**, 205 (1977).

60 S.H. Lin, And W.Y. Lin, *Environ. Technol.* **15**, 299 (1994).

61 J. De la Noue, and D. Proulx, *Appl. Microbiol. Biotechnol.* **29**, 292 (1988).

62 L.C. Rai, and N. Mallick, *World J. Microbiol. Biotechnol.* **8**, 110 (1992).

63 N. Mallick, and L.C. Rai, *World J. Microbiol. Biotechnol.* **9**, 196 (1993).

64 Y.H. Lee, C.W. Lee, and H.N. Chang, *Appl. Microbiol. and Biotechnol.* **30**, 141 (1989).

65 K. Sonomoto, M.M. Hoq, A. Tanaka, and S. Fukui, *Applied and Environmental Microbiology.* **45**, 436 (1983).

66 M.H. Kim, S.B. Lee, D.Y. Ryu, and E.T. Reese, *Enzyme and Microbial technol.* **4,** 99 (1982).

67 P. Chevalier, and J. De la Noue, *Enzyme Microb. Technol.* **7,** 621 (1985 a).

68 P. Chevalier, and J. de la Noue, *Biotechnol. Lett.* **7**, 395 (1985 b).

69 Y. Pouliot, G. Buelna, and J. de la Noue, Procedings of the Renewable Energy Conference 86 "Solar Energy Society of Canada Inc. and the Biomass Energy Institute Inc., Winnipeg, (1986).

70 P.K. Robinson, A.L. Dainty, K.H. Goulding, I. Simpkin, and M.D. Trevan *Enzyme and Microbial Technol.* **7**, 212 (1985).

71 N. Mallick, and L.C. Rai *World J. Microbiol. Biotechnol.* **10**, 439 (1994).

72 J. R. Cumming, and G. J. Taylor, in, Stress Responses in Plants: Adaptation and Aclimation Mechanisms eds. R.G. Alscher, and J.R. Cumming, Wiley-Liss, Inc (1990).

73 D.A. Thurman, in, Effects of Heavy Metal Pollution on Plants ed. Applied Science Publishers, N.W. Lepp, London, England (1981).

74 A. Brookes, J.C. Collin, and D. A. Thurman, *J. Plant Nutr.* **3**, 695 (1981).

75 A. Brune, W. Urbach, and K.J. Dietz, *Plant Cell Environ.* **17**, 153 (1994).

76 K.L. Davies, M.S. Davies, and D. Francis, *Plant Cell Environ..* **14**, 381 (1978).

77 R.F.M. Van Stevenick, M.E. Van Stevenick, D.R. Fernando, W.J. Horst, and H. Marchner, *J. Plant Physiol.* **131**, 247 (1987).

78 R.F.M. Van Stevenick, M.E. Van Stevenick, A.J. Wells, and D.R. Fernando, *J. Plant Physiol.* **137**, 140 (1990).

79 R.F.M. Van Stevenick, M.E. Van Stevenick, and D.R. Fernando, *Plant Soil.* **146**, 271 (1992).

80 W.E. Rauser, *Ann. Rev. Biochem.* **59**, 61 (1990).

81 J.C. Steffens, *Ann.Rev. Plant Physiol. Mol. Biol.* **41,** 553 (1990).

82 D.E. Salt, and W.E. Rauser, *Plant Physiol.* **107**, 1293 (1995).

83 D.E. Salt, and G.J. Wagner, *Biol. Chem..* **268**, 11297 (1993).

84 J. A. de Knecht, M. Van Dillen, P.L.M. Koevoets, H. Schat, J.A.C. Verkleji and W.H.O. Ernst, *Plant Physiol.* **104**, 225 (1994).
85 D.M. Speiser, S.L. Abrahamson, G. Banuelos, and D.W. Ow, *Plant Physiol.* **99**, 817 (1992).
86 P.B.AN. Kumar, V. Dushenkov, H. Motto, and I. Raskin, *Environ. Sci. Technol.* **29**, 1232 (1995).
87 V. Dushenkov, P.B.A.N. Kumar, H. Motto, and I. Raskin, *Environ. Sci. Technol.* **29**, 1239 (1995).
88 T. A. Anderson, E. A. Guthrie and B. T. Walton. *Environ.Sci. Technol.* **27**, 2630 (1993).
89 R.A.H. Smith, and A.D. Bradshaw, *Jornal of Applied Ecology.* **16**, 595 (1979).
90 S. Misra, and L. Gedamu, *Theor. Appl. Genet.* **78**, 161 (1989).
91 D.A. Hammer, ed. *Constructed Wetlands for Wastewater Treatment: Muncipal Industrial and Agricultural,* Lewis Publishers (1989).
92 P. F. Cooper, and B. C. Findlater, *Constructed Wetlands in Water Pollution Control.* Pergamon Press (1990).
93 G.A. Moshiri, ed. *Constructed Wetlands for Water Quality Improvement.* Lewis Publishers (1993).
94 K. Seidel, in : *Biological Control of Water Pollution,* eds. J. Tourbier, and R.W. Pierson, University of Pennsylvania Press (1976).
95 J.B. Robinson, and E. Nieboer, *Lichenologis* **15**, 81 (1983).
96 A. O. Summers, *Annual Reviews of Microbiology.* **40**, 607 (1986).
97 Shen Yun-Fen, A.L. Buikema Jr, W.H. Yongue Jr., J.R. Pratt, and J. Cairns Jr. *J. of protozoology.* **33**, 146 (1986).
98 M. J. Bale, J. C. Fry and M. J. Day. *Applied and Environmental Microbiology.* **54**, 972 (1988).

CHANGES OF SOME SEDIMENTS PROPERTIES IN THE EUTROPHYING LAKES

M V MARTINOVA

Institute of Water Problems, Novo-Basmannaja 10, 107078, Moscow, Russia

1 Introduction

The considerable part of matter and energy taken by lakes accumulates in the sediments. This accumulation can be considered as a result of self purification processes of the water column from surplus matter. The self purification from the dissolved nutrients originates by means of the organic matters (OM) production and sedimentation. After transformation in sediments the part of OM turns into dissolved and gaseous compounds and releases from sediments.

In the chain, the inflow of nutrients in the lake, production of OM^- sedimentation-release from the bottom takes place. The sediments are the stabilization factor, since the flux from the bottom is less than inflow and sedimentation. It is shown that increase of the nutrient concentration (Primarily of P compounds) in the lake's water plays the leading role in the eutrophication. In eutrophying lakes as sedimentation of OM occurs, the release of nutrients from sediments increases. The ratio of external over internal P-load drops from 14 (external P^{-1} gm^2y^{-1}) to 2(internal P^{-3} gm^2 y^{-1})[1]. The stabilization effect of sediments decreases as in the de-eutrophying lakes sediments are often destabilizing factor, since P releases from the bottom can be more than external loading.

The presence of nutrient flux from sediments evidences the non-equilibrium of the transformation processes in sediments. The intensity of OM transformation depends on its concentration and composition. The higher amount of P and N creates the higher gradients and diffusion fluxes. In sediment, higher microbial quantity considerably increases transformation activity. In the eutrophying lakes content of OM and nutrient element increases in the sediments. Non-equilibrium of sediment too increases. The rise in productivity causes the eutrophycation of sediments.

There are some processes of sediment's stabilization. The P compounds are sorbed stronger than N. Chemosorption of P by Fe (OOH) and freshly precipitated carbonate in hard water lakes, reduces P flux from sediments.

Table 1 : Main Characteristics of Narochan lakes.

Index	Lake Naroch	Lake Myastro	Lake Batorin
Area of water surface, km^2	76.6	13.1	6.3
Depth, m :			
average	9.0	5.4	3.0
maximum	24.8	11.3	5.5
of sampling	16.0	8.0	5.0
Hydraulic retention time, years	10.0	2.5	1.0
Primary production, $g\ C\ m^{-2}\ y^{-1}$			
phytoplankton	66	202	258
macrophytes	14	32	13
periphyton	53	18	3
Accumulation of $CaCO_3$, $g\ m^{-2}\ y^{-1}$	11	32	21
Labile Iran in sediments, mg/100g dry seed.	24	29	45
C/N in sediments	13.4	14.5	12.5

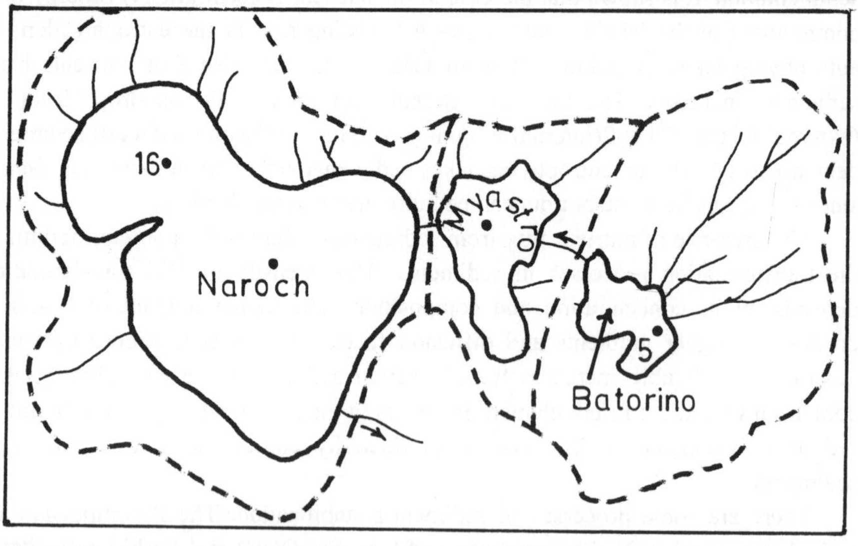

Figure 1 : Index map of Narochansk lakes. Dots indicate sample locations and depth in meter.

The transformation process of sediments changes the chemical composition. In eutrophic lakes, the changes are more intense than one in oligo and mesotrophic. In the present article some correlations for analysis of changes in sediment's property of eutrophying lakes have been discussed.

2 Materials and Methods

The data represents the investigations of the Narochan lakes (Belarus) where I worked with colleagues in 1986-88[2,3,4]. All the measurements were made monthly during 1.5 years at one station in each lakes. The Lakes (Batorin, Naroch and Myastro, Figure 1) are interconnected by channels and are at different states of eutrophication (Table 1).

In terms of physico-chemical properties, sediments and water of three Narochan lakes are similar. The principal differences are the higher content of labile iron in Batorin sediments and the higher rate of $CaCO_3$ accumulation in Myastro lake. Mean values of parameter used are shown in table 2.

Table 2 : Characteristic of the bottom water and sediments.

Variable	Index	Means*			Standard deviations*		
		I	II	III	I	II	III
Solid phase of sediments							
Water content, % wet wt	w	97.2	93.9	95.3	0.8	1.3	2.5
Organic matter, % dry wt	OM	43.3	32.0	39.7	8.6	1.3	4.4
Total particle phosphorus, mg/100g dry wt	FPP	160	160	163	26	11	17
Inorganic particle phosphorus, mg/100g dry wt	IPP	109	91	96	37	10	11
Nonapatite phosphorus,							

mg/100g dry wt	P_{nmn}	50	29	34	21	5	11
Apatite phosphorus, mg/100g dry wt	P_{np}	36	30	24	5	4	4
Residual phosphorus, mg/100g dry wt	P_{rap}	23	32	38	15	6	11
Interstitial water							
Total dissolved phosphorus, layer 0-2 cm., μg l^{-1}	TDP_{0-2}	118	106	119	51	65	72
Total dissolved phosphorus, layer 2-5 cm., μg l^{-1}	TDP_{2-5}	135	82	102	80	29	92
Inorganic dissolved phos- phorus, layer 0-2 cm., μg l^{-1}	$0\text{-}P_{0-2}$	64	36	79	54	22	20
Inorganic dissolved phos- phorus, layer 2-5 cm., μg l^{-1}	$0\text{-}P_{2-5}$	80	30	42	75	16	40
Total dissolved nitrogen, layer 0-2 cm., mg l^{-1}	TDN_{0-2}	5.8	4.7	7.6	2.6	1.4	3.5
Total dissolved nitrogen, layer 2-5 cm., mg l^{-1}	TDN_{2-5}	6.3	5.0	9.1	2.0	1.9	3.4
Ammonium nitrogen, layer 0-2 cm., mg l^{-1}	$NH_{3\ 0-2}$	2.5	1.9	4.9	1.1	0.9	1.9
Ammonium nitrogen, layer 2-5 cm., mg l^{-1}	$NH_{3\ 2-5}$	2.9	2.4	6.8	1.3	1.1	2.3

Bottom water (layer 0-10 cm)							
Temperature, °C	t	6.8	8.1	8.6	5.2	6.0	6.6
Dissolved oxygen, mg l^{-1}	DO	8.5	8.0	6.8	2.4	1.8	2.2
Difference of oxygen Concentrations between surface and bottom water, mg l^{-1}	DDO	3.3	3.1	2.6	3.3	3.1	3.1
Total dissolved phosphorus, μg l^{-1}	TDP	21	24	26	8	17	11
Inorganic dissolved phosphorus, μg l^{-1}	0-P	7	6	9	7	6	14
Total dissolved nitrogen, mg l^{-1}	TDN	1.24	1.44	2.14	0.50	0.60	0.95
Ammonium nitrogen, mg l^{-1}	NH_3	0.20	0.19	0.51	0.15	0.15	0.33
Oxygen consumption by the bottom water, mg l^{-1}	WOD	536	404	507	429	463	365
Oxygen consumption by sediments, mg m^{-2} d^{-1}	SOD	184	177	124	141	141	123
Release of N from sediments, mg m^{-2} d^{-1}	f N	25.7	31.0	32.9	10.8	30.5	12.5
Release of P from sediments, mg m^{-2} d^{-1}	f P	1.70	0.59	0.29	1.00	0.37	0.66

* I - Lake Naroch; II - Lake Myastro; III - Lake Batorin

Interstitial water was derived by centrifugation at 10,000 rpm for 15 minutes and was filtered through a 0.4 μm Nucleopore membrane filter. Bottom water was also filtered. Standard methods were used for analysis.

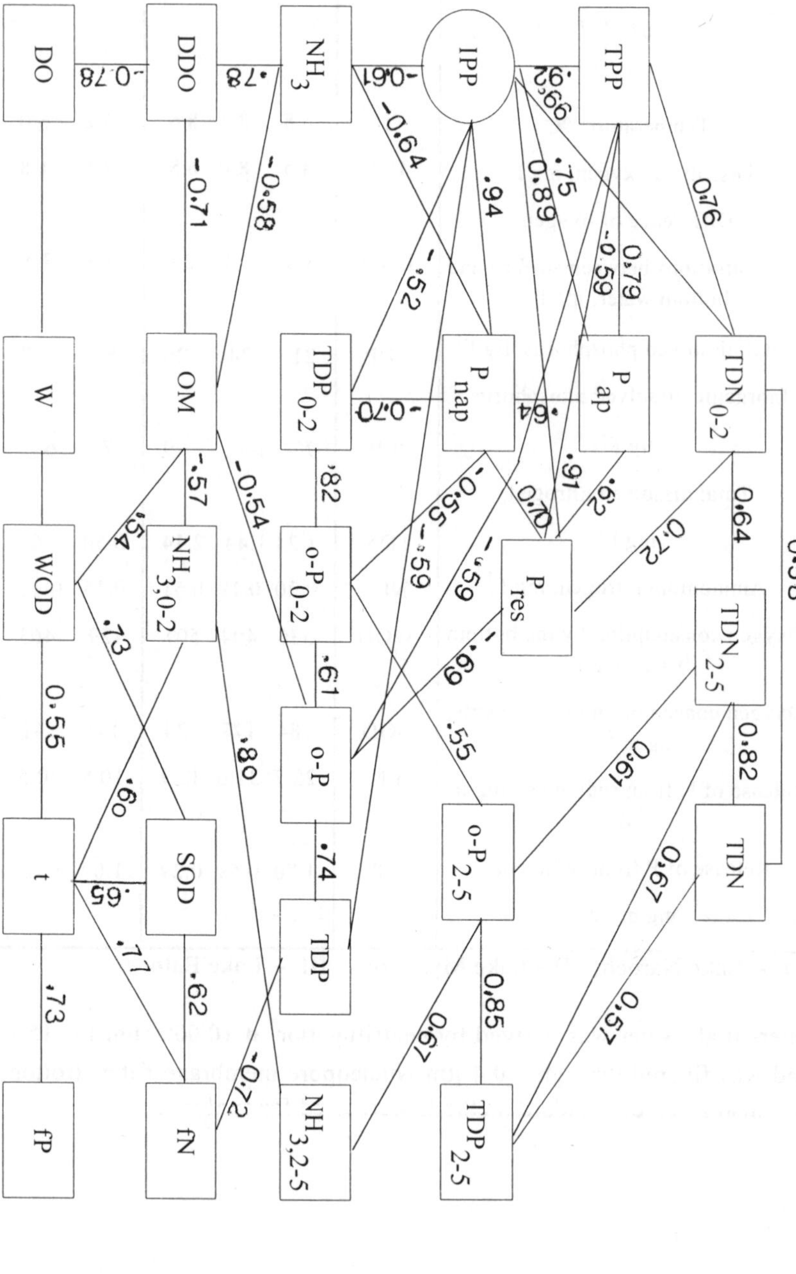

Figure 2 : Some connections in the zone of interactions between water and sediments, Lake Naroch.
Number indicate co-rrelation coefficient.

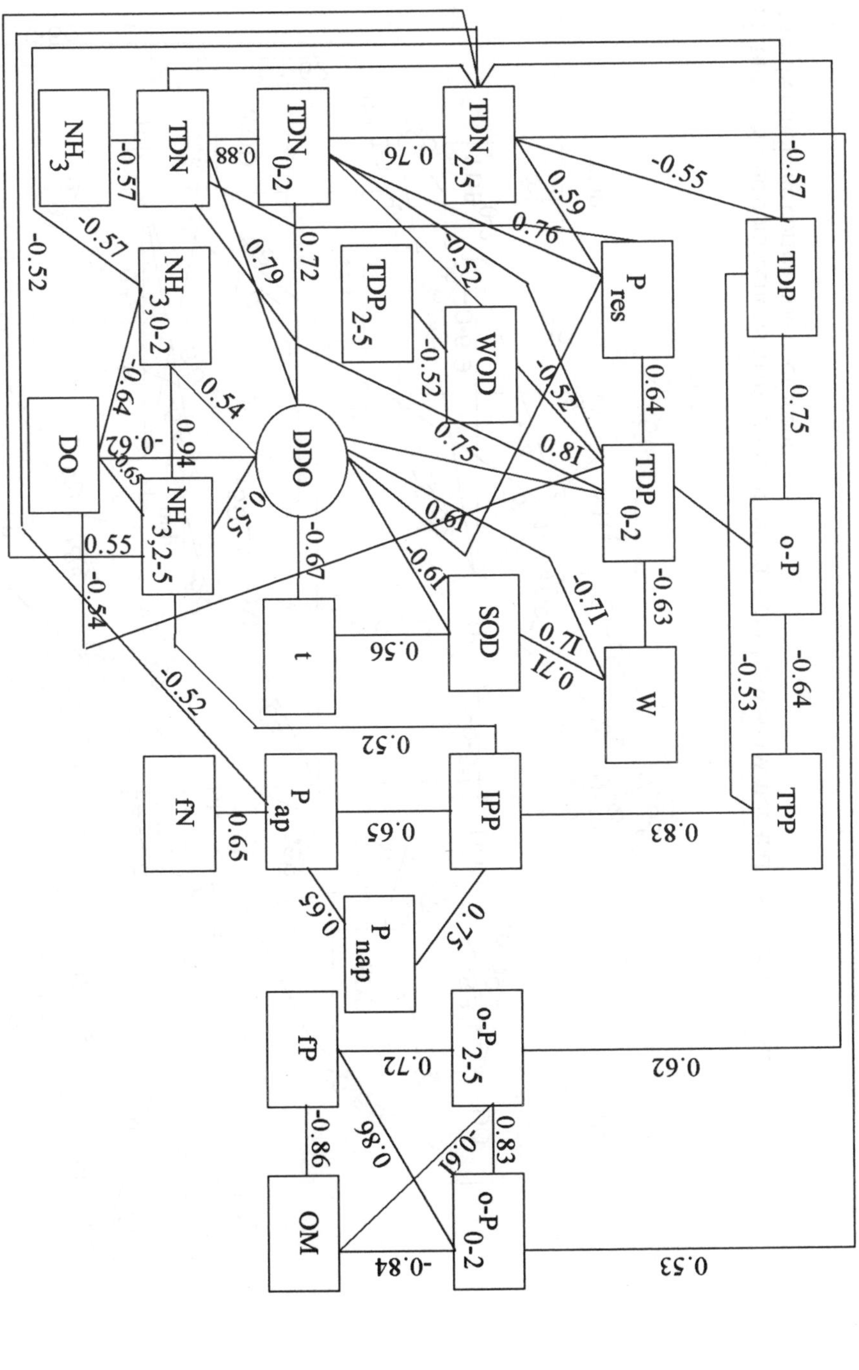

Figure 3 : Some connections in the zone of interactions between water and sediments, Lake Myastro.
Number indicate correlation coefficient

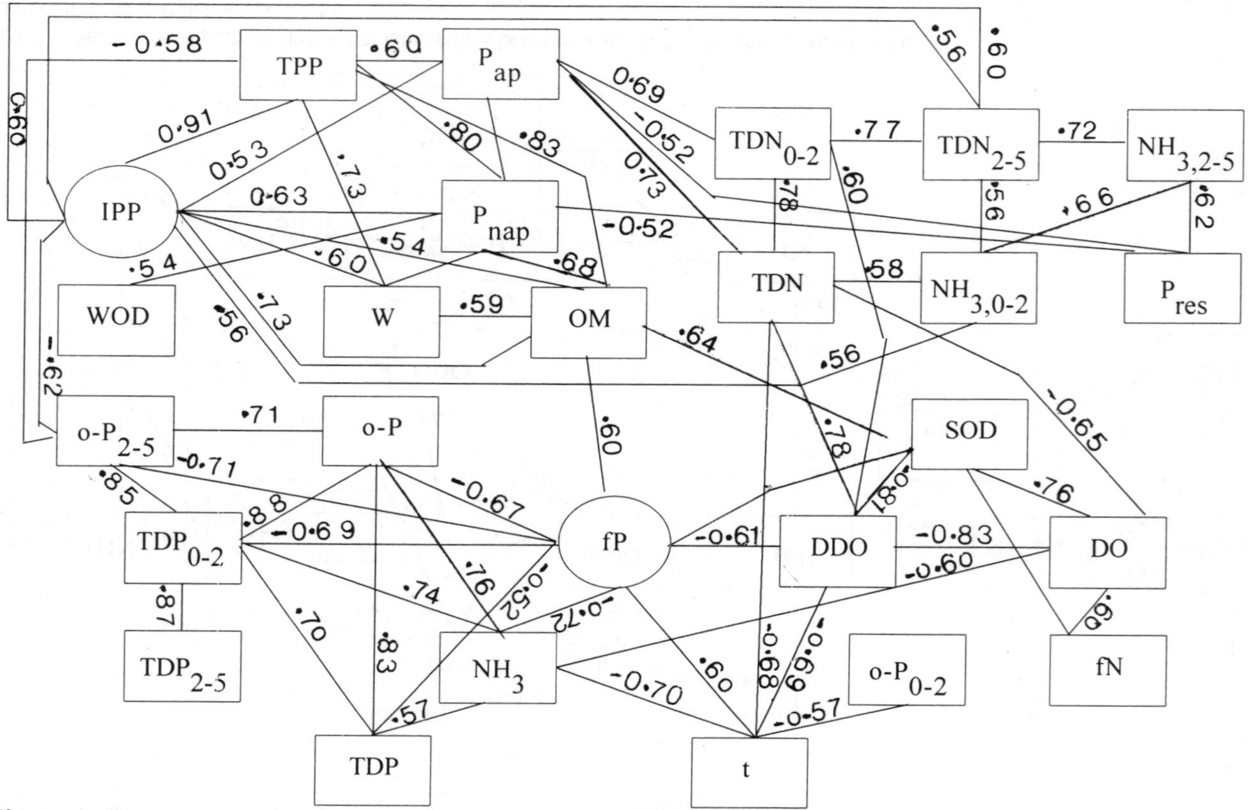

Figure 4 : Some connections in the zone of interactions between water and sediments, Lake Batorin. Number indicate correlation coefficent

3 Discussion

Interactions between water and sediments and their connections for different lakes have been shown schematically in Figure 2,3 and 4. The schemes differ in number of relationships and their ways. Successive connections are prevalent in Naroch lake while cross connections predominates in lake Batorin. The Myastro lake has intermediate situation. Naroch and Myastro schemes have 102 connections, while Batorin scheme has 139 connections. The scheme of Naroch and Myastro lake (Figure 2 and 3) have only 4 parameters with the number of relationships 6-9 whereas Batorin scheme has 14 parameters with the relationships number 6-9. Thus Batorin sediments have more rigid nutrient structure. In the centre of schemes are IPP (inorganic particle phosphorus), (Figure 2), DDO (Figure 3) and IPP, fP (Figure 4). Thus in sediments of eutrophying lake as the rigidity of nutrients increases, the number of their centres also increases. The most increasing number of connection in Narochan lake have DO, fP, W, OM, TDP, OP-2, OP 2-5, NH_3.

The deterioration of oxygen regime is one of the first sign of eutrophication. The Narochan lakes have successful oxygen regime. The minimum oxygen concentration in the bottom water is 4.6 mg l^{-1} for lake Naroch, 4.9 mg l^{-1} for lake Myastro and 3.2 mg l^{-1} for lake Batorin. The relative oxygen consumption is less in the more productive lake Batorin. That is, the increase of the productivity is accompanied by the decrease of the sediments participation in the oxygen consumption. The reverse dependence, the proportion between oxygen consumption rates by bottom water and by sediments from the oxygen concentration in the water, confirms it (Figure 5a, c). In lake Myastro (Figure 5b) it is not so, because station of the samples is in the zone of water current.

The maximum oxygen consumption by the sediments of eutrophying lakes is displayed in the province of the lower temperature. The maximum oxygen demand by the sediments found in lake Naroch at 14-18° C (June). In Myastro at 12-14° C (May) and in lake Batorin at 6-9° C (October). It may be regarded as the result of an impact on the bacterial sediments cenosis by temperature[2]. Waksmann and Renn[5] showed that in rich nutrient media the task of bacterial connections do develop at the lower temperature. It is evident in more eutrophic lake Batorin. The increase of lakes trophic status is accompanied by the increase of factors number regulating the oxygen consumption by sediments. A statistically significant connection of oxygen

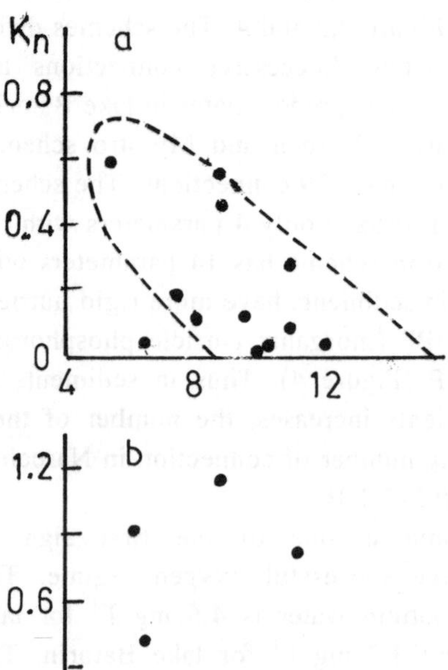

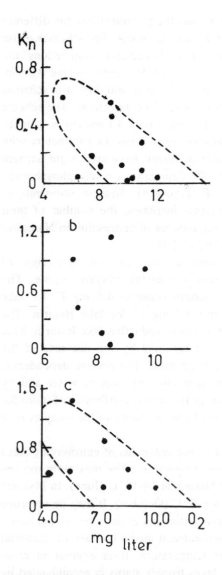

Figure 5 : Effects of oxygen concentration on interrelationship between oxygen consumption rates in bottom water and in sediments, K_n of lakes Naroch (a), Myastro, (b), Batorin(c).

consumption rate exists in each of the lake with different factors of the medium. In the Naroch lake these factors included temperature of the medium and oxygen demand intensity in the water (Figure2) In the Myastro lake the difference in oxygen concentration is due to temperature difference between surface and bottom water layer (Figure 3). In the Batorin lake also oxygen content and organic matter content differs in surface and bottom layers (Figure 4).

The dependence of SOD (oxygen consumption by sediments) upon WOD (oxygen consumed by the bottom water) in the Naroch lake is replaced by the concentration of SOD with fP (release of Phosphorus from sediments) in the Batorin lake. That is the minor deterioration of oxygen conditions in the bottom water increases the influence of SOD on the release of P from sediments.

In the eutrophying lakes if the concentration gradient rises, the rate of nutrient diffusion in sediment also increases. Therefore the role of properties influencing the permeability are more important, In the Naroch sediments the moisture (W) don't bind up with other parameters (Figure 2) In the Batorin lake moisture has 4 connections. The value of OM (organic matter) content (and OM sediment) gets more significance in lake Batorin since the process of its transformation becomes more rapid. The connection of OM with P forms of the sediments indicates the increase content of authigenic P forms (Figure 4). The evidence of relationship between P nap (nonapatite P) and P ap (apatite P) (Figure 2-4) evidence indicates authigenic origin of some part of the inorganic phosphorus.

Preliminary estimations show that the accumulation of the authigenic forms increases from lake Naroch to lake Batorin. The P nap depends on organic P while the relationship between organic P and P ap is independent. This indicates that the chief pathway of the authigenic P ap formation in the sediments is via sorption of phosphates of the interstitial water, where their concentration is controlled by P nap. The connection between content of P ap and carbonates has the correlation coefficient of 0.9. Since the authigenic P ap is less labile its share in the calcareous sediments ($CaCO_3$ content is 13-20%) is some what more than authigenic P nap. In Narochan lakes is is >90% for P ap and ˉ80% for P nap.[3]

Since the value P nap/ P ap in Myastro sediment is less than one in Naroch and Batorin sediments, the rate of the accumulation of $CaCO_3$ (and of P ap) in the Myastro lake is higher (Table 1 and 2). The centre (DDO) of

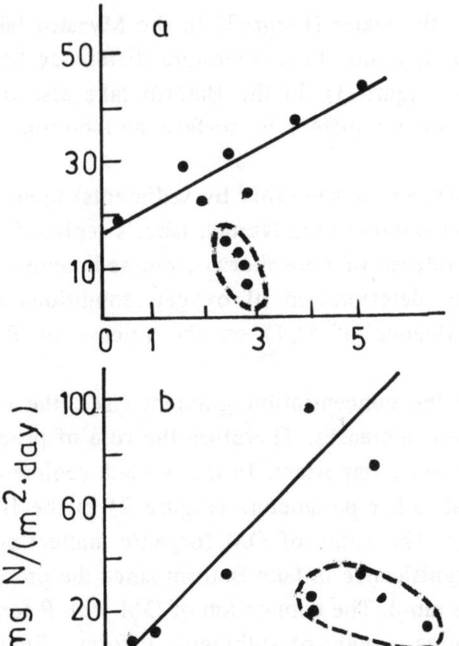

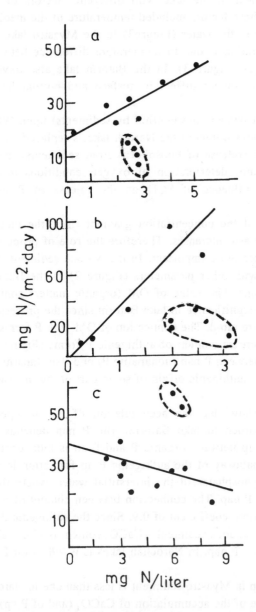

Figure 6 : The flux of N from the bottom as a function of the concentration of NH_3 in the interstitial water of lakes Naroch (a), Myastro (b), Batorin (c).

Myastro lake's scheme (Figure 3) is the indication of the waters dynamic activity that influences upon intensity of the carbonate formation. This process has the most important significance of nutrient transformation (Table 2 and Figure 3).

In the eutrophying lakes dependence of P res (residual phosphorus) from P ap and P nap changes from direct to reverse. Usually inorganic P compounds remain after extractions of Fe and Ca bound phosphate is understood as organic P. The connection between P res and N forms in the sediments (Figure 2-4) shows the part of P res is bound with humic substances. Perhaps P res is humic bound phosphate and Fe- phytate in the organic P fraction of sediments.[6] The increase in TDP (total dissolved phosphorus) relation number is the result of nutrient accumulation in the bottom water of the eutrophying lakes. As it is shown[4] the zone of P accumulation in the eutrophying lake extends gradually from solid phase and interstitial water to the bottom water. The lakes bubbling with nutrients from the sediments shows the release of P from deeper layers of the silt.

Table 3 : Components of N and P balances in sediments of Narochan lakes.

Variable	Lake Naroch	Lake Myastro	Lake Batorin
In surface layers (4 mm) of sediments			
C/N	6.3	8.2	8.7
C/P	94	98	113
Total mineralisation of OM in sediments			
$g\ C\ m^{-2}\ y^{-1}$	51.5	45.7	55.5
$g\ N\ m^{-2}\ y^{-1}$	8.2	5.6	6.4
$g\ P\ m^{-2}\ y^{-1}$	0.55	0.47	0.50
Fluxes from bottom.			
$g\ N\ m^{-2}\ y^{-1}$	9.5	11.2	12.3
$g\ P\ m^{-2}\ y^{-1}$	0.40	0.26	0.19
Sorption of P			
$g\ P\ m^{-2}\ y^{-1}$	0.15	0.21	0.31
% of mineralisation	27	45	62
Desorption of N			
$g\ m^{-2}\ y^{-1}$	1.3	5.6	5.9
% of mineralisation	14	50	48

In the eutrophying lake release of NH_3 from rediment is more. The direct relationship between N release from the bottom and NH_3 concentration in the interstitial water changes by reverse relationship (Figure 6). the deeper layers of the sediment involves in these releases. The upper millimetres of the sediment play less important role for the release of dissolved nitrogen in the water. I used an earlier proposed method[1] for the estimation of the contribution of the deeper layers of the sediments, in the N- fluxes from the bottom of three lakes (Table 3).

In the lake Naroch this contribution is less than one in two other lakes. In the Batorin lake there is a connective diffusion. It is created by bubbles from bottom. Therefore, greater N flux from the bottom corresponds to less of N-concentration in the interstitial waters.

References

1. M. V. Martinova, *Hydrobiologia,* **252**, 1 (1993).
2. M. V. Martinova, T. V. Zhukova, and E.P. Zhukov, *Water Resources,* **2**, 123 (In Russ) (1991 a).
3. M. V. Martinova, T. V. Zhukova, and E. P. Zhukov, *Water Resources,* **3**, 90 (In Russ) (1991 b).
4. M. V. Martinova, T. V. Zhukova, and E. P. Zhukov, *Water Resources,* **3**, 73 (In Russ) (1992).
5. S. A. Waksmann, and Ch.E. Renn, *Biol. Bull.,* **70**, 472 (1986).
6. H. L. Golterman, *Hydroblologla* **315**, 35 (1995).

RESTORING DEGRADED ECOSYSTEMS IN THE LAURENTIAN GREAT LAKES

JOHN H. HARTIG

International Joint Commission
100 Ouellette Ave.
Windsor, Ontario, Canada N9A 6T3

1 Introduction

The Laurentian Great Lakes are a unique global feature, spanning the international border between Canada and the United States. The Great Lakes contain 20% of the world's surface freshwater. Within the drainage basin of the Great Lakes can be found some of the world's largest centers of commerce, industry, and urban development, and unfortunately, human-induced degradation.

Degradation of the Great Lakes Basin Ecosystem is the result of over a century of human population growth and economic/industrial development. The combined effects of population growth and development are manifested most in nearshore embayments, harbors, and river mouths. Currently, there are 42 degraded areas of the Great Lakes or Areas of Concern where one or more beneficial uses are impaired (Figure 1).

For each of the Areas of Concern, a remedial action plan (RAP) is being developed to identify and implement key actions needed to restore beneficial uses. The concept of RAPs originated from a 1985 recommendation of the International Joint Commission's Great Lakes Water Quality Board (IJC 1985). The Board found that despite implementation of regulatory pollution control programmes, a number of beneficial uses were not being restored and recommended that comprehensive and systematic RAPs be developed and implemented to restore all beneficial uses in Areas of Concern. The 1987 Protocol amending the Canada-United States Great Lakes Water Quality Agreement[1] formalized the RAP programme and explicitly defined Areas of Concern as specific geographic areas that fail to meet the general or specific objectives of the Agreement where such failure has caused or is likely to cause impairment of beneficial use or impairment of the area's ability to support aquatic life. Impairment of beneficial use means a change in the chemical, physical, or biological integrity of the Great Lakes ecosystem sufficient to cause any one of 14 use impairments (Table 1). Currently, 36 of the 42 Areas of Concern have restrictions on fish and wildlife consumption, and restrictions on dredging activities, 35 have degradation of benthos, 34 have loss of fish and wildlife habitat,

30 have degradation of fish and wildlife populations, 25 have degradation of aesthetics, 24 have beach closings, 21 have eutrophication or undesirable algae, and 20 have fish tumors or other deformities (Table 1)[2].

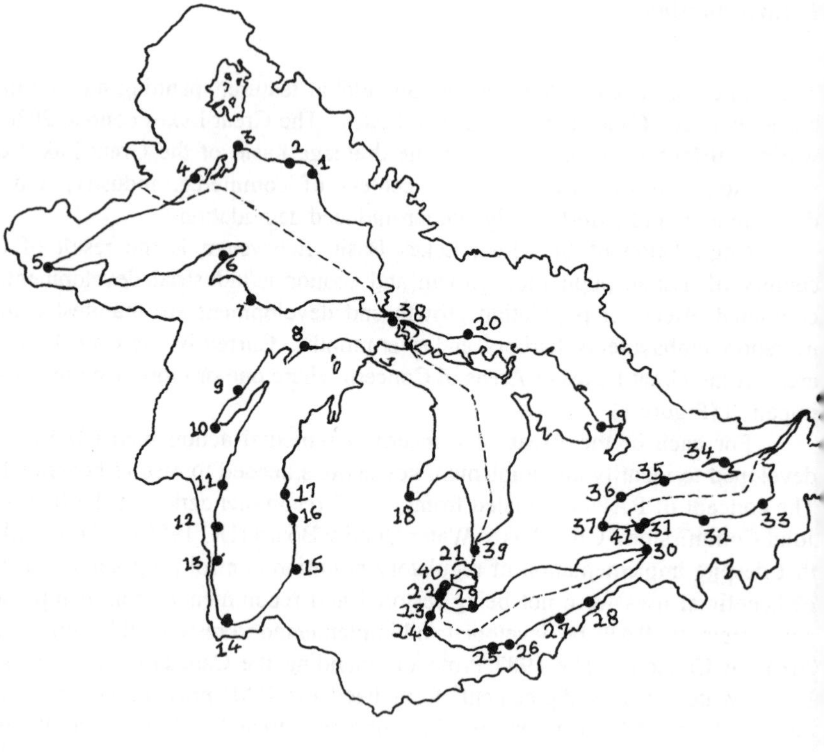

Figure 1 : Fortytwo areas of concern identified in the Great lakes basin ecosystem.

Lake Superior	**Lake Erie**
1. Peninsula Harbour	21. Clinton River
2. Jackfish Bay	22. Rouge River
3. Nipigon Bay	23. River Raisin
4. Thunder Bay	24. Maumee River
5. St. Louis Bay/R	25. Balck River
6. Torch Lake	26. Cuyahoga River
7. Deer Lake	27. Ashtabula River
Carp Crek/R	28. Presque Isle Bay
Lake Michigan	29. Wheatley Harbour
8. Manistique River	30. Buffalo River
9. Menominee River	31. Eighteen Mile Creek
10. Fox River	32. Rochester Embayment
11. Sheboygan River	33. Oswego River
12. Milwaukee Estuary	34. Bay of Quinte
13. Waukegan Harbour	35. Port Hope
14. Grand calumet River/Indiana	36. Metro Toronto
Harbour canal	37. Hamilton Harbour
15. Kalamazoo River	**Connecting Channels**
16. Muskegon Lake	38. St. Marys River
17. White Lake	39. St. Clair River
18. Haginaw River/ Saginaw Bay	40. Detroit River
19. Severn Sound	41. Niagara River
20. Spanish River	42. St. Lawrence River

Table 1. A summary of the extent of use impairments in Great Lakes Areas of Concern.

Use Impairment	Number of areas of concern with the impaired use (n=42)
Restrictions on fish or wildlife consumption	36 (86%)
Tainting of fish and wildlife flavour	4 (10%)
Degradation of fish and wildlife populations	30 (71%)
Fish tumors or other deformities	20 (48%)
Bird or animal deformities or reproductive problems	14 (33%)
Degradation of benthos	35 (83%)
Restrictions on dredging activities	36 (86%)
Eutrophication or undesirable algae	21 (50%)
Restrictions on drinking water consumption, or taste and odor problems	12 (29%)
Beach closings	24 (57%)
Degradation of aesthetics	25 (60%)
Added costs to agriculture or industry	7 (17%)
Degradation of phytoplankton or zooplankton populations	10 (24%)
Loss of fish and wildlife habitat.	34 (81%)

Annex 2 of the Great Lakes Water Quality Agreement states that RAPs shall embody a systematic and comprehensive ecosystem approach to restoring and protecting uses in Areas of Concern[1]. In addition, the Agreement states that the Parties (i.e., federal governments), in cooperation with State and Provincial Governments, shall ensure that the public is consulted in all actions undertaken pursuant to RAPs. The RAP programme has been described as an experiment in adaptive, environmental management where flexible, locally-designed, ecosystem approaches are being used to build the capacity to restore beneficial uses[3]. This chapter provides a summary of the lessons learned in over ten years of the Great

Lakes RAP Programme and presents a case study of Collingwood Harbour, the first Area of Concern to restore all uses.

2 Use Of An Ecosystem Approach To Develop And Implement RAPs

An ecosystem approach accounts for the interrelationships among water, land, air, and all living things, including humans; and involves all user groups in comprehensive management[4]. The Great Lakes Water Quality Agreement has outlined an iterative, action-oriented, three-staged process for practical application of an ecosystem approach in use restoration. Stage 1 involves reaching agreement on a comprehensive problem definition (Table 1) and identification of causes and sources of degradation. Stage 2 evaluates and identifies remedial and preventive actions, and generates an action agenda (i.e., secures commitments and endorsements, with implementation schedules). Implementation is an ongoing process and does not wait for completion of the planning process. Stage 3 provides data and information to confirm use restoration.

Currently, 40 of the 42 Areas of Concern have either a stakeholder group, coordinating committee, or comparable RAP institutional structure broadly representative of environmental, economic, and societal interests in Areas of Concern. These RAP institutional structures are established to: help implement an ecosystem approach; ensure broad-based public participation; help coordinate and facilitate RAP development, audit RAP implementation, and track progress; and help build the capacity to restore beneficial uses.

A key concept in the RAP process is accountability for action[2]. This is established through open sharing of information, clear definition of problems (including identification and quantification of indicators to be used in measuring when the desired state is reached)[5], identification of causes, agreement on actions needed, and identification of who is responsible for taking action. From this foundation, the responsible institutions and individuals can be held accountable for progress.

RAPs require cooperative learning that involves stakeholders working in teams to accomplish a common goal under conditions that involve positive interdependence (i.e., all stakeholders cooperate to complete a task) and individual and group accountability (i.e., each stakeholder is accountable for the final outcome). For RAPs to be successful, they must:

* be cleanup- and prevention-driven, and not document-driven;
* make existing Programmes and statutes work;
* cut through bureaucracy;
* elevate the priority of local issues;

* ensure strong community-based planning processes;
* streamline the critical path to use restoration; and
* be an affirming process.

Based on a basin-wide review of progress in the Great Lakes RAP Programme, RAP processes are most effective if they are mission-driven (i.e., a focus on ecosystem results and restoring uses) and not rule-driven[6]. Successful RAP processes empower institutional structures to pursue their mission of restoring impaired uses. Empowerment of RAP institutional structures can be demonstrated by: a focus on watersheds or other naturally-defined boundaries to address upstream causes and sources, and obtain commitments from within the watershed for implementation; an inclusive and shared decision-making process; clear responsibility and sufficient authority to pursue the mission; an ability to secure and pool resources according to priorities for action using nonprofit organizations or other creative mechanisms; flexibility and continuity in order to achieve an agreed-upon road map to use restoration; commitment to broad-based education and public outreach; and an open and iterative RAP process that strives for continuous improvement[6].

3 Restoration Of Collingwood Harbour

Collingwood Harbour is situated on the south shore of Nottawasaga Bay, which constitutes the southern extension of Lake Huron's Georgian Bay. The harbor is surrounded by the Town of Collingwood, which has a permanent resident population of approximately 15,000. The harbor includes a wetland complex, a wastewater treatment plant outfall, marina, grain terminal, and a former shipyards. Impaired uses in the harbor in the early 1980s included: eutrophication or undesirable algae; degraded fish and wildlife populations; loss of fish and wildlife habitat; and restrictions on fish and wildlife consumption[7].

The Public Advisory Committee for the Collingwood Harbour RAP was created in 1988 to help develop and implement the RAP. It played a key leadership role in raising public awareness of RAP activities, forming partnerships to implement remedial and preventive actions, and recruiting thousands of local volunteers to participate in activities ranging from water course rehabilitation to a pollution prevention Programme. In addition, an intergovernmental RAP Team provided technical support to the RAP process.

Public participation was extensive in the Collingwood Harbour RAP process. In addition to its consultative role in establishing goals and beneficial uses for Collingwood Harbour, the Public Advisory Committee evaluated and recommended a remedial and preventive strategy to achieve those goals and uses. The Stage 1

RAP was completed in 1989, the Stage 2 RAP completed in 1992, and the Stage 3 RAP completed in 1994.

In 1995, Collingwood Harbour became the first Great Lakes Area of Concern to restore all impaired beneficial uses and be delisted as a Great Lakes Area of Concern. This is significant to note because governments, working in true partnership with the stakeholders of Collingwood Harbour, were able to successfully use a locally-designed ecosystem approach to restore degraded fish and wildlife populations, remediate contaminated sediments, abate cultural eutrophication problems, and rehabilitate fish and wildlife habitat. The Collingwood Harbour RAP and its Public Advisory Committee provide a practical example of successful use of science, research, technology, and an ecosystem approach in a multi-stakeholder process to improve ecosystem quality and restore and maintain both human and non-human uses of a degraded watershed.

Public involvement in the Collingwood Harbour RAP process and community-based partnerships were essential to generate the political will and public support for the necessary remedial and preventive actions. Indeed, Beeker et al.[8] have shown that a critical factor for acceptance of a RAP by all involved is to structure the process in such a way as to create a sense of ownership of the plan by the participants, who would be the very businesses, state and local agencies, and citizens who would have to carry out its recommendations.

One of the major accomplishments of the Collingwood Harbour RAP process was applying an innovative technology to optimize phosphorus removal at the Collingwood Harbour Water Pollution Control Plant. Phosphorus removal at the plant was enhanced by multi-point chemical addition (i.e., alum), automated alum dosage control, and raw and biological solids inventory control through computer-assisted operation[9]. Automation of the chemical feed system to pace chemical addition rates to the phosphorus loading to the plant was approached by use of on-line phosphorus analyzers in conjunction with automatic pump control. Dual-point chemical addition was incorporated by implementing alum addition to the primary clarifiers in conjunction with the current practice of adding alum to the secondary treatment system. Monitoring data have shown that the Collingwood Harbour Water Pollution Control Plant is capable of achieving high efficiency phosphorus removal and low effluent phosphorus concentrations. This project optimized phosphorus removal at a cost savings of $9.4 million, representing a "win-win" situation for the environment and economy.

Another major accomplishment of the Collingwood Harbour RAP process was undertaking a sediment removal (8,000 m^3) project which was the first demonstration of this technology (i.e., Pneuma Pump System) in North America[9]. The Pneuma Pump System is based on a principle of using static water head and compressed air inside cylinders. Each of three cylinders is rapidly filled with slurry by counter pressure due to a hydrostatic head and induced vacuum. When one cylinder is filled, compressed air acts as a piston and the slurry is then forced through a check valve to the discharge pipeline. The pump has no rotating parts or mechanisms in contact with the sediment, minimizing sediment resuspension problems. Contaminated sediment was removed from the harbor slips and transported through a pipeline to a confined disposal facility 1.2 km away. A silt curtain located at the north end of each slip was used to confine any possible particle resuspension due to unforeseen dredging complications. Full-scale sediment cleanup was achieved for $1.36 million. Monitoring data demonstrated achievement of biologically-based cleanup targets, while minimizing the problems of secondary pollution caused by disturbance at the sediment-water interface.

Other major accomplishments of the Collingwood Harbour RAP process included:

* using stormwater management practices to control erosion and enhance habitat in Black Ash Creek;
* implementing a harbour habitat rehabilitation programme;
* undertaking a comprehensive pollution prevention programme entitled "The Greening of Collingwood" which targeted all sectors, including resource conservation and hazardous product alternatives;
* implementing Watershed Planning;
* linking the RAP to land-use planning by incorporating the RAP principles into the Official Plan for the Town of Collingwood; and
* developing an environmental playground called "Enviropark" and incorporating educational materials on the RAP and the environment into local school programmes.

The Collingwood Harbour RAP process represents practical application of ecosystem approach theory, effective community-based partnerships, effective use of science, research, and engineering, and successful multi-stakeholder management of a watershed. The stakeholders in Collingwood have now launched a preservation initiative to ensure that the harbor remains clean and useable.

Conclusions

Considerable progress is being made in re-orienting Great Lakes decision-makers to a more inclusive and holistic RAP process that accounts for linkages, shares decision-making power, achieves local ownership, and focuses on ecosystem results. Although most Areas of Concern have experienced ecosystem improvements or use restoration as a result of the RAP process, continued progress in many Areas of Concern may well depend on overcoming technical, institutional, and financial obstacles to remediation of contaminated sediments, and on gaining public acceptance of proposed remedial actions.

The development and implementation of RAPs represent the first opportunity, on a broad and practical scale, to implement an ecosystem approach consistent with the long-term goal of sustainable development (i.e., development that meets the needs of the present generation without compromising the ability of future generations to meet their own needs). Both an ecosystem approach in RAPs and sustainable development recognize the fundamental roles and interrelationships of economy, society, and environment in sustaining the quality of ecosystems. Table 2 presents a list of critical elements to help environmental managers guide efforts toward ecosystem-based management of watersheds[10]. In a time when financial resources for environmental programmes are becoming scarce, such an approach has proven an effective mechanism to coordinate programmes and harness and leverage resources.

Governments in Canada and the United States are learning that all stakeholders have something to offer and can play significant roles in RAPs. Through government and community-based partnerships, RAPs are attempting to overcome environmental decision-making gridlock by developing a coordinated, multi-stakeholder response to restoring impaired beneficial uses in Areas of Concern. In essence, RAPs are building the capacity (i.e., human, scientific, technological, organizational, institutional, and resource capabilities) to restore and sustain beneficial uses in Great Lakes Areas of Concern (Table 3). Sustaining the RAP process will require continuing public involvement, ensuring long-term commitment to research and monitoring, achieving effective communication and cooperation, creatively acquiring resource commitments, and building a record of success. Both short- and long-term milestones must be celebrated. Examples of milestones include: government management actions; remedial and preventive actions by sources; changes in discharge quality; reductions in contaminant loadings; changes in air/water/sediment concentrations; reductions in bioaccumulation rates; biological recovery; use restoration; and improved suitability for human use of resources.

Most countries throughout world are working to rehabilitate or enhance degraded regions or to preserve high quality regions. Although not all lessons learned in the Great Lakes RAP process are transferable to other areas around the world, many of them could provide a practical starting point for others interested in applying an ecosystem approach in management or in improving upon existing management Programmes. Some of the key lessons from the Great Lakes RAP process include:

* there are positive effects from opening up the decision-making process through meaningful public participation (roles and responsibilities should be clearly defined and understood);

Table 2. A list of critical elements to help government managers guide efforts toward ecosystem-based management of watersheds[10].

* Adopt watershed/bioregion as primary unit for management
* Develop partnership agreement or other mechanism for cooperative, multi-stakeholder management and ensure commitment of leaders
* Identify and empower an "umbrella" watershed organization for coordination
* Develop long-term vision, goals, and quantitative targets for "desired future state" of ecosystem
* Reach agreement on a set of principles to guide decision-making process
* Ensure all planning processes in watershed acknowledge vision, goals, quantitative targets, and principles
* Establish geographical information system (GIS) and decision support system capability in watershed organization
* Compile data and information for input into GIS and ensure strong commitment to research and monitoring to understand ecosystem and fill knowledge and data gaps
* Set priorities that target major causes of ecosystem health risks, evaluate remedial and preventive options, implement preferred actions, and monitor effectiveness in an iterative fashion (i.e., adaptive management)
* Ensure full costs and benefits (i.e., economic, societal, environmental) are assessed for each project in watershed
* Consolidate capital budgets and pool resources, as necessary, to move high priority projects forward
* Create the framework and conditions for private sector involvement and capitalize on its enterprise, initiative, creativity, and capability for investment
* Utilize market forces and economic incentives to achieve ecosystem objectives
* Commit to public, state-of-the-environment and economy reporting every 2-5 years to measure and celebrate ecosystem progress, and to measure stakeholder satisfaction
* Ensure ecosystem-based education and human resource development throughout process

Table 3. Ways and means to enhance and strengthen the capacity to restore uses in Great Lakes Areas of Concern[11].

Areas of Capacity Building	Ways and Means to Build Capacity to Restore Uses
Human elements and strategies	* Empowerment * Decentralized authority * Long-term vision/mission driven * Partnerships/coalitions * Shared decision-making * Integrated planning processes * Broad-based participation * Human resource development * Public outreach and education * Securing commitments and endorsements
Tools and techniques	* Pollution prevention * Best management practices * Habitat rehabilitation and enhancement * Sediment remediation * Hazardous waste site remediation * Urban redevelopment and design
Management support systems	* Watershed planning and management * Performance measures * Geographical information systems and decision-support systems * Monitoring and research * Information sharing

* an integrated, ecosystem approach can result in eliminating overlap and increasing efficiency;
* priority should be placed on clarifying and reaching agreement on problems (e.g., the 14 use impairments in Table 1 provide a template for problem definition) and priorities;
* emphasis should be placed on planning cooperatively and sharing responsibilities for delivery of Programmes;
* strong efforts should be made to build partnerships (don't be afraid to experiment with new processes and approaches);

* governments should provide resources to initiate and catalyze multi-stakeholder decision-making processes (these investments of resources often result in substantial leveraging of nongovernmental and private sector resources);
* coupling of research and management has proven time and again to be cost- and ecosystem-effective; and
* emphasis should be place on measuring and celebrating progress and striving for continuous improvement.

References

1. United States and Canada. International Joint Commission, Windsor, Ontario, Canada, (1987).
2. J. H. Hartig, and N. L. Law, *Environmental Management* **18,** 855 (1994a).
3. J. H. Hartig, and M. A. Zarull. eds. University of Michigan Press, Ann Arbor, Michigan, U.S.A., (1992).
4. J. H. Hartig, and J.R. Vallentyne. *Ambio* **18**, 423 (1994).
5. J. H. Hartig, , M. A. Zarull, G. Mikol, T.B. Reynoldson, V. A. Harris, R.G. Randall, and V. W. Cairns. in, Proceedings of the 67th Annual Conference and Exposition of the Water Environment Federation, Water Environment Federation, Alexandria, Virginia, U.S.A. (1994).
6. J. H. Hartig, and N. L. Law. eds. Environment Canada, Toronto, Ontario, Canada; U.S. Environmental Protection Agency, Chicago, Illinois, U.S.A., (1994b).
7. Environment Ontario and Environment Canada. Collingwood Harbour Remedial Action Plan: Stage 1. Toronto, Ontario, Canada, (1989).
8. J. Beeker, G. Studen, and L. Stumpe, in, *Surface and Groundwater Quality,* eds. A. A. Jennings and N. E. Spangenberg, *American Water Resources Association*, TPS-0101, 29-41 (1991).
9. G. Krantzberg, and E. Houghton, Ontario Ministry of Enviornment and Energy, Toronto, Ontario, Canada, (1994).
10. T. Coape-Arnold, S. Crockard, K. Fuller, J. E. Gannon, S. Gerritson, J. H. Hartig, N. L. Law, G. Mikol, K. Mills, L. New, A. Richardson, K. Seidel, and M.A. Zarull, U.S. Environmental Protection Agency, Chicago, Illinois; International Joint Commission, Windsor, Ontario, Canada (1995).
11. J. H. Hartig, N. L. Law, D. Epstein, K. Fuller, J. Letterhos, and G. Krantzberg. *Int. J. Sustain. Dev. World Ecol.* **2,** 1 (1995).

ECOLOGY OF EUTROPHIC LAKE UDAISAGAR AND RESTORATION

B.C. RANA

Department of Biosciences, Sardar Patel University, Vallabh vidyanagar 388120 Gujarat, India

1 Introduction

Eutrophication of natural water bodies is one of the significant water quality problem of modern time. The eutrophication is a kind of water pollution which is a consequence of nutrient enrichment. The nutrient enrichment leads to mass production of algae and undesirable biotic changes. Such a progressive deterioration of water body is referred as eutrophication. According to Vollenweider[1]. Eutrophication is seen as increasing enrichment of surface waters with plant nutrients, with all its applied consequences. Increasing nutrient supply results in accelerated productivity affecting all compartments of a limnetic system[2]. Numerous factors are involved in eutrophication. In general geochemical, climatic, hydrological, limnological, catchment characters, properties of water and anthropogenic alterations determine its productivity[3,4]. Phosphorus is generally considered as the decisive limiting factor of primary production in oligotrophic and mesotrophic ecosystems[5]. Nitrogen has been reported as a limiting factor mainly in lentic waters, with high amounts of available phosphorus.[6] Increased nutrients lead to growth of algae and cyanobacterial bloom.[8,9] Major quantitative and qualitative changes in algal, bacterial and macrophytic populations are involved in the process of eutrophication. The appearance of algal bloom, discolouration of water, fish kills and emergence of marshland are the common time-lag phenomenon of a eutrophic process.

Conservation and management of our aquatic ecosystems particularly of those used in drinking and industrial water supplies are essential. Udaipur a prominent city of Rajasthan, a city of lakes, has large number of lakes in and around the city. These lakes are not only the beauty spots and tourist attraction but are also the source of drinking, irrigation and industrial uses. These lakes have been damaged to the extent that all measures to restore will fail in near future if no action is taken. Lake Udaisagar approximately 18 km away from Udaipur city is one of the important source of industrial and irrigation water supply. This lake is the main source of water supply to the Zinc Smelter Debari, a factory of Hindustan Zinc Ltd. created at a cost of more than 100 crore rupees. Although the lake is away from urban dwellings and direct anthropogenic stress of Udaipur city, is getting polluted by agricultural runoff and drainage of river Ayad which carries the domestic and industrial wastes of Udaipur city. The deterioration of water quality, at times due to heavy algal bloom, interrupts industrial water supply and causes a great loss to the industry.

In the present chapter limnology of the lake Udaisagar with reference to eutrophication has been discussed. The restoration measures suggested by the research team and measures taken by authorities have also been discussed. The restoration approaches suggested by various limnologists for restoring eutrophic lakes have also been summarised. Ecology of large number of Indian lakes have been studied, only few of them have been referred here[10-19].

2 Lake Udaisagar

Lake Udaisagar is historically important. It is also a source of industrial, irrigation and drinking water supply. Lake is situated in the South East of Udaipur at a distance of 18 km between 24° 32' 30" to 24° 35' N latitude and 73° 48' to 73° 49' 30" E longitude. The water in the lake is stored by making a dam of 900 feet long and 180 feet broad between two hills on the river Berach. The lake is triangular in shape and covers an area of 7.25 hact. however, it swells to 41.0 hact. during rainy season. The depth of the lake on dam side is 15.25 meter. The average rainfall of this area is about 500mm.

2.1 Study Sites

In the present study six study sites were selected (Figure 1) Out of the six study sites, three were located near deep water zone, fourth was a shallow water zone, far off from the first three sites. This area is occupied by large number of hydrophytes and intensive agriculture is done during winter and summer months. Study site five and six represent inlet point of river Ayad and river water near industrial area respectively.

3 Materials and Methods

Observations on various hydrobiological parameters of Udaisagar lake were carried out from 1985 onwards up to 1992. From the selected study sites samples were

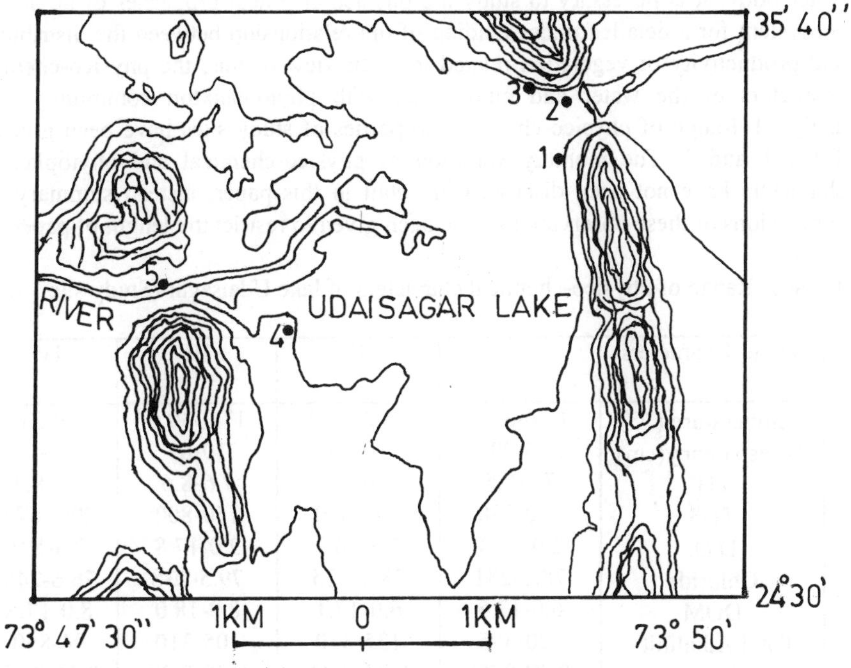

Figure 1 : Map of the study site indicating sampling points (1 to 5).

collected intensively for first two years and subsequently seasonal data were collected. The following physico-chemical parameters were analysed as per Standard Method[20]: Temperature, pH, transparency, total dissolved solids (TDS), free CO_2, dissolved oxygen (DO), chlorides, total alkalinity (TA), dissolved organic matter (DOM), chemical oxygen demand (COD), nitrate and phosphates. Samples for the investigation of phytoplankton were collected from all the six study sites along with water samples using plankton net of bolting silk with 50 mesh/cm. The concentrated plankton samples were preserved in Lugol's solution. Plankton were identified using standard monographs and published work. Wet land community of littoral zone was studied by phytosociological methods. Soil and bottom mud samples were collected and analysed for nutrient contents.

4 Observations and Conclusions

The spectrum of life in a lake is closely related to its physico-chemical conditions. There goes on an intricate pattern of environment-community inter-actions in a water body. It is necessary to study the physico-chemical properties of an aquatic ecosystem for a detailed understanding of the relationship between the distribution and productivity of vegetation related to it. In view of this, the physico-chemical characters of the water and mud along with phytoplankton community were analysed. Range of physico-chemical properties of study sites have been given in Table 1 and 2. The monthly variation of physico-chemical and phytoplankton characters have not been discussed in detail in this paper, instead summary and conclusions of these observations have been given to restrict the length of paper.

Table 1 Range of physico-chemical characters of lake Udaisagar (Study site 1 to 4)

Water Characters	I	II	III	IV
Temperature (°C)	17.0-30.0	16.0-30.0	17.0-30.0	16.9-31.7
Transparency(cm)	10-90	10-85	10-85	--
pH	7.7-8.5	7.6-8.5	7.6-8.6	7.2-8.4
TDS	150-750	200-600	150-800	200.-675
D.O.	2.0-17.4	2.4-16.2	2.0-17.8	2.4-8.9
Chloride	78.1-231	78.1-235	79.5-238	76.6-248
DOM	6.0-17.5	6.0-17.1	8.0-18.0	8.0-17.5
Total Alkalinity	120-395	135-330	105-310	94-850
Nitrate	0.28-0.89	0.33-0.94	0.38-0.99	0.36-0.89
Phosphate	0.29-0.97	0.36-1.2	0.44-0.92	0.22-1.2

All results except Temperature, pH and transparency are in ppm

Table 2 Range of physico-chemical characters of river Ayad (Study site 5 and 6)

Water Characters	V	VI
Temperature (°C)	16.0-29.0	20.5-37.5
pH	7.0-8.1	7.0-8.2
TDS	1600-5500	1700-2400
D.O.	0.8-6.0	1.8-7.6
Chloride	184-350	160-320
Nitrate	3.5-5.0	2.9-5.4
Phosphate	2.5-5.2	2.3-7.7
COD	205-850	240-860

All results except Temperature, and pH are in ppm

4.1 Physico-chemical Characters

The physico-chemical characters of study sites one to four have been given in Table-1. The temperature of surface water of lake and inflow sewage of river Ayad is more or less in conformity with atmospheric temperature. Sacchi disc transparency was less in summer and rainy months because of thickness of algal bloom, turbidity due to suspended solids and intensity of light. High values of TDS (> 100 ppm) indicate eutrophic nature of lake water[21]. Water of all the sampling sites were alkaline in nature throughout the study period. High DO content was observed in October, November and December months while rest of the months showed poor DO content of lake water. Poor DO content may be due to intense microbial activities in the bottom and water column. More DOM observed during summer and monsoon may be due to decomposition of organic matter and inflow of sewage rich Ayad river and agricultural runoff. High concentration of chloride speaks of organic pollution due to sewage discharge. The concentration of chloride above 60 ppm indicate heavy pollution. The total alkalinity of lake water was due to bicarbonate as the value of carbonates were very less. The higher values of alkalinity (> 50 ppm) indicate that water of the lake is hard water. The high concentration of nitrate specially during rainy season further supports the conclusion that the source of pollution is domestic sewage and agricultural runoff. The values of phosphate was also very high during rainy season in the lake. The physico-chemical characters of study site four and five (river Ayad) indicated high nutrient statues with reference to nitrate, phosphate, chlorides, TDS COD alkalinity while DO was negligible (Table 2). In general it could be stated that inflow of river Ayad having high nutritional and pollution values play an important role in eutrophying the lake Udaisagar.

Table 3: Soil and sediment composition of Lake Udaisagar (% dry soil).

Characters	Soil	Sediment
Soil moisture	6.5-25	6.2-28
Water holding capacity	10.3-30.4	11.8-33.4
pH	7.1-7.7	7.1-7.9
Chloride	0.01-0.07	0.01-0.09
Bicarbonates	0.05-0.18	0.06-0.19
Nitrogen	0.01-0.06	0.01-0.08
Phosphate	0.001-0.009	0.001-0.008
Organic Carbon	0.02-0.38	0.26-0.43

The soil and bottom mud analysis of the lake (Table 3) indicated high concentrations of various nutrients. It serve as potential reservoir of nutrients and are exchanged with water and affects the composition and abundance of plankton and other biota of water bodies.

4.2 Phytoplankton

Enumeration of phytoplankton of study sites one to four revealed more than 40 taxa belonging to 38 genera (Table-4). Numerical composition of each species was also determined. Chlorophyceae was represented by 18 genera and 18 species and was dominant algal group. Xanthophyceae was most poorly repented having only *Botryococcus braunni*. Cyanophyceae was an important group from the point of waterbloom formation in lake Udaisagar. This group was represented by eight genera and ten species. Bacillanophyceae was represented by nine taxa. *Euglena viridis* and *Phacus longicaude* were the representatives of Euglenophyceae Two species of *Microcystis, M. aeruginora* and *M. flosaquae* were perennial and formed algal bloom during summer and rainy months. The dominance of chlorococcales over poorly represented desmids is an indication of eutrophic nature of the lake. The abundance of *Botryococcus braunni* requiring high concentration of nutrients indicates a high level of pollution in the lake. The abundance of genera like *Chlamydomonas* sp, *Pediastrum simplex, Scenedesmus quadricauda, Pandorina mourn, Chlordla valgaris, Oscillatoria princeps, Microcystis aeruginosa, Merismopedia minima, Nitzschia gracilis, Navicula lanceolata, Synedra ulna, Euglena viridis* clearly indicate the high pollution level of lake Udaisagar. On the contrary study site five and six were represented by less number of taxa (Table-5), many of them were common, however it differs from planktonic forms of the lake. Nygaard's trophic index,[22] for different phytoplanktonic groups were estimated and all the three groups showed eutrophic nature of the lake.

Table 4 : List of phytoplankton in lake Udaisagar (Study site 1,2,3 and 4)

Cyanophyceae :	*Ulothrix* sp
Microcystis aeruginosa	*Zygnema cyanaum*
M. flosaquae	*Spirogyra macrospora*
Oscillatoria princeps	*Spirogyra palludosa*
Oscillatoria spirulinoides	*Mougeotia* sp.
Nostoc sp.	*Closterium lanceolatum*
Anabaena sp.	*Cosmerium renijorme*
Anabaenopsis sp	**Xanthophyceae :**
Merismopedia sp	*Botryococcus braunii*

Phormidium sp	**Bacillariophyceae :**
Chlorophyceae :	*Cyclotella operculata*
Chlamydomonas sp.	*Fragillaria brevistriate*
Pandorina morum	*Navicula lanceolate*
Pediastrum simplex	*Nitzschia gracilis*
Pediastrum duplex	*Synedra ulna*
Chlorella sp.	*Pinnularia viridis*
Scenedesmus quadricauda	*Gomphonema* sp.
Hydrodictyon reticulatum	*Diatoma* sp.
Cladophora sp.	*Stauroneis anceps*
Aphanochaete repens	**Euglenophyceae :**
Oedogonium sp.	*Euglena viridis*
Bulbochaete sp.	*Phacus longicauda*

4.3 Macrophytes

Aquatic macrophytes confined themselves to the littoral zone of the lake and contribute significantly to the total primary production in wet-land ecosystems,[23,24] The phytosociological studies of lake Udaisagar showed the study area four harboured 23 species (Table-6). Out of these, thirteen represented moist bank species, five floating species and the remaining five were submerged species.

Table 5: List of algal of river Ayad (Study site 5 and 6)

Cyanophyceae	**Chlorophyceae**
Merismopedia minima	*Scenedesmus quadricauda*
M. punctata	*Charasiosiphon rivalaris*
Oscillatoria limosa	*Hydrodictyon* sp.
O. subliformis	*Ulothrix zonata*
O. curviceps	*Spirogyra* sp.
Spirulina subsalsa	**Bacillariophyceae**
Anabaena torulosa	*Nitzschia amphibia*
	Navicula exigua

Of the twenty angiospermic species eleven were monocot and nine dicot showing a ratio of 1.2:1. The comparison of the three groups shows that moist bank species were dominant over floating and submerged species. The occurrence of common submerged species *Potamegeton pectinatus* suggests hardwater nature of the lake.

Presence of *Eichhornia crassipes* and *Lemna trisulca* weed complex in shallow region of the lake indicates a serious threat for the lake by these species. The moist bank rooted plants like. *Alternanthera sessilis* and *Phyla nodiflora* can translocate the nutrients from the sediments to water. However a large shallow littoral zone of this lake helps in retaining the wastes brought by river Ayad and restricts its direct entry into the lake. However, during rainy season nutrients and macrophytes are washed into the lake.

Table 6 : List of macrophytes in lake Udaisagar

Moist Bank Species	**Floating :**
Alternanthera sessilis	*Eichhornia crassipes*
Paspalum distachyum	*Lemna trisulca*
Polygonum barbatum	*Wolfia arrhiza*
Trianthema monogyna	*Nymphoides indicum*
Phyla nodijlora	*Azolla pinnate*
Sporobolus diander	
Elipta prostrata	**Submerged**
Heliotropium supinum	*Hydrilla verticillata*
Cyperus alopecurioides	*Najas minor*
Cyanodon dactylam	*Ceratophyllum demersum*
Monniera herpestis	*Potamogeton pectinatus*
Echinochloa crusgalli	*Chara corallina*
Marsilea quadrifolia	

In an attempt to find chemical control of algal bloom laboratory trials were carried out with certain chemicals. Many algicidal and algistatic compounds have been found effective in the control of nuisance algae[25]. The most promising chemical groups having algicidal properties are inorganic salts, organic salts, amines, antibiotics, quinones, substituted hydrocarbons, quaternary ammonium compounds, amide derivatives and phenols. Copper sulphate has been in use for the selective control of algal growth in lakes and reservoirs since early 1900. In the present study also effect of different concentrations of copper sulphate, thorium nitrate and sodium nitride were tried on *Microcystis aeruginosa* under laboratory conditions. Copper sulphate at 1.0 ppm was found effective while other compounds were not effective at lower concentrations. However, use of any chemical for such a large water reservoir is not practical as it is cost effective and affects not only the target organism but many associated organisms also.[26] In recent studies Parker et al[27] showed that more than 5 mm Potassium in water bodies inhibit *Microcystis* bloom and the ratio of K: Na play an important role in bloom formation.

4.4 Conclusion

Almost all the parameters of physical, chemical and planktonic analysis of lake Udaisagar revealed that lake has reached to the eutrophic level. A composite rating index using large number of parameters of such lake may be useful not only in confirming its eutrophic nature but it can provide the trend of eutrophication useful for the control of particular parameter.[28]

It may be concluded from the present study that lake Udaisagar is highly eutrophic and polluted. This is also indicated by the regular occurrence of thick algal bloom of *M. aeruginosa*. Frequent mass mortality of fish have been observed. Zinc Smelting Factory, who draws its water from this lake had to close for many times incurring a very heavy monetary losses.

5 Restoration

The long standing complaints of local residents regarding poor quality of water, foul smelling due to water blooms and fish mortality, and many disorders of grazing cattles drinking this water was never taken seriously by the local or responsible authorities. Attention was drawn only when Zinc Smelting factory had to close frequently due to interruption of water supply. Efforts were made to improve the quality of lake water by following suggestions and their implementations.

1. Diversion of river Ayad water entering the lake
2. Water treatment plant for sewage and industrial wastes of Udaipur city
3. The catchment area and corridor management for the control of soil erosion and agricultural ran off
4. Mechanical harvest of agal bloom
5. Ban on agriculture in the littoral zone

The initial measures taken for diverting river Ayad water entering the lake, tree plantation, soil conservation and mechanical harvest of algal bloom helped to improve the lake water quality to certain extent. The record of last four years have revealed no mass mortality of fish, no thick algal bloom and no closure of zinc smelter due to water supply. However, our regular observations have revealed that still the eutrophic nature of lake water has not changed much. The occurrence of permanent agal bloom of *M.aeruginosa* and other planktonic algae, high amount of phosphate, nitrate, chlorides and dissolved organic matter are in the eutrophic side. The greatest potential of lake nutrients are its bottom mud. However the continued restoration efforts will be going to improve the water quality of lake Udaisagar a good sign.

Many suggestions and methods have been developed to reverse the process of eutrophication of water bodies[29]. Many of such methods suggested by large number of limnologists have been summarized below[30].

1. *Ecosystem development* : The strategies of ecosystem development provide a most important basis for evaluating measures of water quality control and lake preservation[31]. Table-7 lists some examples of measures and how grossly they affect the water quality. It is evident that water pollution control consists not only of water treatment, but many other physical and biological means of water management. Proper management of catchment and corridor land is significant. Reduction in residence time of nutrients produces desirable effect .

2. *Phosphorus elimination* : The first and most important measure is to reduce the nutrient loading of inland waters. Phosphorus removal from point source discharges is more practicable for inland waters than nitrogen elimination. The reduction of phosphorus loading reduces the total reserve of phosphorus in lake and thus reduces the ultimate capacity of biomass production. Since the potential productivity depends on the rate of supply of soluble and assumable phosphorus the reduction of phosphorus loading can prevent algal bloom. More complete phosphorus removal can be accomplished by treating wastes with lime, iron or aluminium salts leading to chemical precipitation of phosphate. High degree of phosphorus elimination can be achieved by pH adjustment.

3. *Limiting the amount of nutrients entering the lake* : The nutrients limitation can be achieved by (a) Waste water treatment upto tertiary level (b) Removal of phosphorus and nitrogen at their source. (c) The diversion of nutrients rich effluent from water bodies (d) Dilution of nutrients in the lake by controlled addition of water low in nutrients.

4. *Reduction in the amount of nutrients solubilization* in water through microbial decomposition of bottom sediments. This can be achieved by artificially planting an inert layer which covers bottom sediment.

5. *Cultivation and harvesting of waste grown algae* : This enables to remove inorganic nutrients. Pollution tolerant and mutagenic strains with high gleaning capacity and protein contents can be used for this purpose.

6. *Removal of dissolved nutrients by chemical or physical methods* : Employing physico-chemical methods of pre, simultanious and post precipitation as high as 90 percent phosphorus can be removed, Other chemicals can also be removed by inorganic floculants and synthetic organic compounds. Zirconium oxychloride is a promising chemical as it can precipitate up to 95 percent phosphate[32] .

7. *By encouraging the natural food web* : So as to graze on algae by aquatic biota and subsequent harvest of fish.

8. *Self cleaning ability of water* bodies can be increased by selectively favouring the decomposers.

9. *Nitrogen removal* from the effluent can be achieved by (i) biological nitrification and denitrification (ii) air stripping of ammonia from an alkalized waste water (iii) ion exchange (Iv) electrodialysis (v) reverse osmosis.

Table 7 : Lake restoration by promoting ecological stability.

1. P-R balance restoration
(i) By reducing waste input
(ii) harvesting biomass
(iii) reducing residence time
(iv) Mixing synthesis and decomposition
(v) Fish management
(vi) Aeration
2. Conservative land management
(i) By reforestation
(ii) Restricting monoculture productivity
(iii) Maintaining zones adjacent to open waters with low productivity
(iv) Controlling soil erosion
3. Enhancement of biological complexity
(i) By establishment of ecological niches
(ii) Seeding diverse population and recirculating certain organisms
(iii) Maintaining relatively high biomass
(iv) Maintaining stratification
(v) Selective harvesting
(vi) Maintaining high chemical buffer intensity

10. *Algicide and herbicide* treatment may provide temporary relief from excessive growth of algae and macrophytes but in long run they may prove harmful as they reduce the diversity and stability of aquatic system.

11. *Aquaculture:* The high fertility of waste waters can be manipulated for harvesting the useful aqua products.

12. *Aquatic macrophytes* like *Eichhornia crasipes* if harvested can be used for removing the inorganic nurtient loading from the water bodies.

13. *Management of wetlands* play critical role in the maintenance and improvement of water quality due to natural processes of arresting the pollutants, discharged through direct releases, removal of sediments, and production of oxygen.

References

1. R. A. Vollenweider, Rep.Org. Econ. Coop Dev. DAS/CSI/68 27 Paris (1968).
2. R. G. Wetzel, *Limnology* Saunders Company Philadelphia USA (1975).
3. C. N. Sawyer, *J. Water Poll. Control Fed.* **38**, 737 (1966).
4. R. A. Vollenweider, *Z. Wasser Abwasserforsch* **12**, 10 (1979).
5. W. Stumm and E. Stumm- Zollinger, in *Water Pollution Microbiology,* ed. R Michell, Wiley Intersciences, NY (1972).
6. J. J. Goering, in *Water Polllution Microbiology,* ed. R Michell, Wiley Intersciences, NY (1972).
7. A. Claesson and S. O. Ryding, *Prog. Water Tech.* **8**, 291 (1977).
8. E. A. Thomas, in *Environmental Phosphorus Hand Book* ed. E. J. Griffith *et al.*, Wiley and Sons, NY (1973).
9. V. Venkateswarlu, *Ind. J. Bot.* **4**, 31 (1980).
10. D. P. Zutshi and K. K. Vass, in *Aquatic Weeds in S. E. Asia* ed. C. K. Varshney and J. Rzoska, Dr. W. Junk Publishers, The Hague (1976).
11. V. Kaul, *Int. J. Ecol. Environ. Sci.* **3**, 29 (1977).
12. S. M. Das and J. Pandey, *Indian J. Ecol.* **5**, 7 (1978).
13. K. K. Vass, *Hydrobiologia,* **68**, 9 (1980).
14. G. R. Hegde and S.G. Bharati, *Phykos,* **23**, 71 (1984).
15. T. C. Khatri, *Environ. & Ecol.,* **5**, 71 (1987).
16. D. P. Zutshi, and A. U. Khan, *Ind. J. Environ. Hlth.* **30**, 348 (1988).
17. K. R. Reddy, R. Jessup and P. S. C. Rao, *Hydrobiologia,* **159**, 177 (1988).
18. B. C. Rana, R. P. Vijayvergia and Y. D. Tiagi, *Indian J. Ecol.*, **18**, 161 (1991).
19. B. C. Rana, J. I. Nirmalkumar and S. S. Sreenivas, in *Algal Ecology: An overview,* ed. A. N. Kargupta and E. N. Siddique, International Book Distribution, Dehradun, India (1995).
20. A P H A *Standard Methods for The Examination of Water and Wastewater,* American Public Health Association NY (1976).
21. A. M. Beeton, *Limnol. Oceanogr.* **10**, 240 (1965).
22. R. K. Trivedi and P. K. Goel, *Chemical and Biological Methods for Water Pollution Studies,* Enviro. Publication Karad, India (1984).
23. J. I. NirmalKumar, Rita Nirmalkumar and B. C. Rana, *J. Bomb. Nat. Hist. Soc.,* **88**, 210 (1991).
24. B. C. Rana, J. I. NirmalKumar, *Inter. Jour. Ecol. Environ. Sciences.,* **18**, 195 (1992).

25. G. P. Fitzerald, *Appl. Microbiol,* **7**, 205 (1959).
26. B. C. Rana, J. I. NirmalKumar, *Bull. Environ Cont. Toxicol,* **55**, 104 (1995).
27. D. L. Parker, H. D. Kumar, L. C. Rai and J. B. Singh, in *Algal Biology and Biotechnology* (Symposium), CAS in Botany, Banaras Hindu University, Varanasi (1997).
28. B. C. Rana, J. I. NirmalKumar, *J. Environ. Biol.* **14**, 111 (1993).
29. R. T. Oglesby and W.T. Edmondson, *Jour. Water Poll. Control Fed.* **38**, 1452 (1966).
30. B. C. Rana, in *Environmental Pollution and Health Problems* ed. Rais Akhtar, Ashish Publishing House, NewDelhi (1990).
31. E. P. Odum, *Science,* **164**, 262 (1969).
32. H. D. Kumar and L. C. Rai, *Aqu. Botany,* **4**, 359 (1978).

BIODEGRADATION OF HALOGENATED COMPOUNDS

J.S.H. TSANG AND Y.B. HO

Department of Botany, The University of Hong Kong, Hong Kong.

1 Introduction

As a result of advances in industrial technology and the intensification of agricultural practices, various new, xenobiotic, organic and inorganic compounds have been produced. These compounds have been released into the environment in large quantities either deliberately or accidentally. Many of these synthetic and organic compounds may be reacted with the surrounding discharged area abiotically. There are photochemical transformations, nonenzymatic and nonphotochemical modification at or near the surface of water, soil, and vegetation, but most of these processes rarely mineralize the organic molecules and many of these reactions did not even alter the characteristics of the original molecules. Other than abiotic means biological processes are also reactive. Such biological processes include the interaction of plants and to a certain extent, animals in natural or man-modified environments. However, such biological interaction which involves organismal enzymes as catalysts may cause different types of unwanted changes in the chemicals. Some of these transformation may break down the chemicals but some may enhance the recalcitrance and even toxicity of the molecules. Nevertheless, most of the biological processes actually occurred in the micro-organisms inhabited in soil, sediment, wastewater, surface and groundwater where the chemicals were released.

There are far too many chemicals being treated as pollutants and in this article we would like to concentrate just on the halogenated compounds. Table 1 shows some of the halogenated compounds being used widely in agriculture and industry and which are also causing environmental problems because of their persistence. And as a result of their resistance to biodegradation these chemicals have accumulated in the environment drastically. The recalcitrance, toxicity and sometimes carcinogenicity of such compounds has led to widespread concern and considerable effort has been directed towards development of biological and non-biological detoxification procedures.

Table 1 : Persistence of some halogenated compounds in soil (Alexander[1]).

Compounds		Persistence (Years)
BHC	Hexachlorocyclohexane 1,2,4,5,6,7,8,8-	16
Chlordane	Octahloro-2,3,3a,4,7,7a-hexahydro-4,7-methanoindane	16
Chlorfenvinphos	2-Chloro-1-(2,4-dichlorophenyl)vinyl diethyl phosphate	4
	1,1,1-Trichloro-2,2-bis(ρ-chlorophenyl) ethane	
DDT	3,6-Dichloro-2-methoxybenzoic acid	21
	1,2,3,4,10,10-Hexachloro-6,7-exoxy-	
Dicamba	1,4,4a,5,6,7,8,8a-endo, exo-1,4:5,8-	4
Dieldrin	dimethanonaphthalene	21
	1,2-Dibromoethane (ethylene dibromide)	
	1,4,5,6,8,8-Heptachloro-3a,4,7,7a-	
EDB	tetrahydro-4,7-methanoindene	19
Heptachlor	γ-1,2,3,4,5,6-Hexachloro-cyclohexane	16
	Polychlorinated dibenzofurans	
Lindane	Polychlorinated dibenzo-ρ-dioxin	21
PDF	Pentachlorophenol	8
PDD	2,4,5-Trichlorophenoxyacetic acid	8
PCP	2,3,6-Trichlorobenzoic acid	5
2,4,5-T		1
2,3,6-TBA		4

The biological methods involve the use of extensive catabolic potential possessed by micro-organisms. An understanding of the evolution of these metabolic pathways may help to engineer strains with improved degradative activities. The biodegradation of haloorganic compounds is of particular interest in this respect for two reasons: (i) Their recalcitrance and toxicity, (ii)

Previous studies have been shown that specific metabolic pathways have been evolved to detoxify/mineralize these compounds.

1.1 Halogenated Compounds and Higher Plants

Although contamination of our environment by anthropogenic halogenated organic compounds, including some very toxic and persistence ones such as pesticides (e.g. DDT, aldrin, chlordane), polychlorinated biphenyls (PCBs), dioxins and chlorinated phenols[2], is a major concern for the scientists, politicians, and members of the public alike, relatively much less attention has been drawn to the natural production of chlorinated compounds. A recent review[3] indicated that some 2,000 halogenated organic compounds are produced and released into the biosphere by higher plants and other living organisms[4,5]. Synthesis of halogenated compounds are readily accomplished by haloperoxidase enzyme in plants[6] and, apparently, a wide range of such compounds including alkaloids[7], steroids[8], and even dioxins[9] can be produced. For example, it has been shown that horse-radish peroxidase converts chlorophenols to both polychlorinated dibenzodioxins (PCDDs) and polychlorinated dibenzofurans (PCDFs)[9]. Thus we now realize that many of the halogenated organic compounds may come from natural, biogenic sources as well.

Plants do accumulate exogenous sources of halogenated organic compounds. The route of entry of these compounds depends on a number of factors, such as the hydrophobicity and volatility of the compounds, the soil conditions and the plant species itself. Generally there are five major pathways by which a plant can accumulate these compounds. They are (a) adsorption onto the root surface, (b) uptake into the root, (c) absorption of volatilized compounds by the shoot, (d) absorption from soil particles that contaminate the aerial plant parts, and (e) adsorption/absorption of the airborne compounds by the shoot[10]. Once taken into the plant, the halogenated compounds may be restricted within the organ, with little or no inter-organ transportation, such as in the carrot[11] and cucumber[10], or conversely, readily translocated between different parts of the plant as in zucchini and pumpkin[10]. Different tissues within the same plant may even have preference to take up different halogenated compounds. For example, carrot-root peel and foliage take up tetrachclorobenzenes efficiently whereas carrot root core absorbs trichlorobenzenes preferentially[11]. Thus if plants are to be used of for the biodegradation of halogenated organic compounds, the mode of uptake, the accumulation and response characteristics of the plant towards these compounds should be known beforehand.

As indicated earlier, many organohalogen compounds are readily biodegraded by microorganisms. However the same may not apply to higher plants and much less is known on their ability to biodegrade these compounds. It was shown that PCBs taken up by pine needless from the atmosphere accumulated and persisted in the epicuticular wax. The PCB concentrations increased with the age of the needless up to 8 years[12] indicating that biodegradation was minimal. Indeed the persistence of these compounds in coniferous trees has led to their use as biomonitors for PCBs years, PCDDs and PCDFs, and DDT[11,13].

Some recent attempts were made in trying to genetically engineer tree species for biodegradation of hazardous wastes and for site restoration. Various tree species were screened for their ability to accumulate and degrade halogenated organic compounds. Two approaches were tried. First was to use *Agrobacterium rhizogenes* to engineer for larger root biomass for the degradation of the pollutants in the rhizosphere[14]. Second was to use DNA constructs of two bacterial genes for the enzyme chlorophenol hydroxylase and chlorocatechol 1,2-dioxygenase to make two separate plant expression vectors, plasmids pDCPH1-GUS and pDIGI. The constructs were then employed to transform three tree species, *Robinia pseudoacacia* (black locust), *Populus* sp. (Hybrid poplar) and *Liquidambar styraciflua* (sweetgum) using *Agrobacterium rhizogenes*. Subsequent root culture and shoot regeneration of these plants indicated that T-DNA was inserted into the plant genome and the two inserted genes expressed[15]. The genetic engineering should prove useful in the utilization of plants to bioremediate polluted sites in future.

2 Biodegradation of Halogenated Aliphatic Compounds

2.1 The Chemistry of Halogenated Aliphatic Compounds

Table 2 outlines the properties of the halogens, which undergo an orderly change in properties according to the positions in Periodic Table except fluorine, which tends to stand apart from the other halogens.

The distinction between fluorine and the other halogens in terms of size and electronic properties was emphasized by Pauling's designation of it as a "super halogen"[16]. It is much smaller than the other halogens and its electronegativity makes the carbon-fluorine bond relatively polar, compared with the carbon-chlorine and carbonbromine bonds. Since iodine has an electronegativity comparable to that of carbon, the carbon-iodine bond would be expected to be the least polar. The carbon-fluorine bond energy is amongst the highest found in natural products, whilst the energies of the other carbon-halogen bonds are comparable to the more common bonds made or broke in intermediary metabolism.

Table 2: Properties of the Halogens and other atoms.

Atom	Atomic Weight	Van der Waals Radius (A°)	Electronegativity	Bond Energy (Kcal/mol) to carbon	to hydrogen
F	18.9984	1.35	4.0	105.4	134.6
Cl	35.453	1.80	3.0	78.5	103.2
Br	79.909	1.95	2.8	65.9	87.5
I	126.9044	2.15	2.5	57.4	71.4
H	1.00797	1.20	-	98.8	104.2
C	12.01115	-	2.5	83.1	98.8

Replacement of hydrogen with a halogen in the alkanoic acids yields derivatives which ionize to a greater extent than the unsubstituted forms. The halogens, being strongly electronegative, tend to attract electrons, thus, the inductive effect of the chlorine makes the chlorinated alkanoic acids stronger acids than the corresponding acetic or propionic acids[17]. The effect of chlorosubstitution on the pK_a values of several acids is shown in Table 3. The inductive effect of chloro-substitutions on acidity decreases rapidly with increasing distance from the carboxylic acid group. For example, compare the acidity of 2- and 3- chloropropionic acid. The effect of other halogens on the magnitude of electron attraction decreases in the order of decreasing electronegativity (Cl>Br>I).

The chlorinated alkanoic acids are water soluble, anionic compounds that do not possess functional groups generally associated with hydrogen bonding. These facts suggest that little or no adsorption of these compounds on soil colloids should occur. Since Dalapon or TCA can exist in the acid form only at very low pH values, kearney *et al.*[18] suggested that the ionic species encountered under most biological conditions will be the anions, dichloropropionate and trichloracetate. They concluded that the appearance of the acid or undissociated form would be extremely rare under most physiological conditions.

Table 3 : Effect of chlorosubstitution on acidity of acetic and propionic acids at 25° C.

Acid	pK_a	Acid	pK_a
Acetic	4.76	Propionic	4.88
Monochloroacetic	2.81	2-chloropropionic	2.80
Dichloroacetic	1.29	3-chloropropionic	4.10
Trichloroacetic	0.08	2,3-chloropropionic	1.71

The chemistry of compounds such as TCA and Dalapon is deceptively simple. The compounds are easily prepared and the products undergo standard reactions. In the laboratory, α-chlorinated acids can undergo dehydrochlorination, yielding (according to reaction conditions) hydroxy acids, amino acids, or cyano acids. In water solutions, these acids decompose at room temperature, for example, TCA forms chloroform and carbon dioxide. The dry Dalapon sodium salt is stable, but aqueous solutions are subject to decomposition to form pyruvic acid and inorganic chloride ions. The reaction does not take place readily under acid conditions and proceeds very slowly at temperatures below 20° C.

Another factor other than the type of substitution that influence the degradation of halogenated aliphatics is the number of substitution on the α-carbon and on adjacent carbon (β-carbon). Table 4 shows the dissociation energies required for some of these compounds.

An increase in α-carbon substitution increases the degradability while a similar increase in substitution on the β-carbon decreases the reactivity of the compound. Finally, substitution on unsaturated carbon molecules, e.g. trichloroethene, are recalcitrant than saturated molecules, e.g. 1, 1, 2-trichloroethane[19].

Table 4: Dissociation energies required for the carbon-halogen bond at 25° C

Compounds	ΔH (kcal/mole)	Compounds	ΔH (kcal/mole)
CH_3F	108.0	CH_2Cl_2	81.0
CH_3Cl	83.4	$CHCl_3$	77.7
CH_3Br	70.6	CCl_4	70.0
CH_3I	55.5	CH_3Br	70.6
CH_3CH_2Cl	81.8	CH_2Br_2	61.0
CH_2CHCl	83.6	CBr_4	50.0

2.2 Degradation of Haloalkanoic Acids

Although the production of synthetic haloorganic compounds is causing environmental concern there are many natural haloalkanoic acids found in the biosphere. Most of them exist as metabolic intermediates of natural complex halogenated compounds [20,21].

2.2.1 2-Haloalkanoic Acids

Microorganisms have been playing a key role in the mineralization of these halogenated alkanoic acids. Hydrolysis is one of the most common mechanism utilized by microorganisms in the degradation of halogenated aliphatic acids. The enzymes catalyzing such reaction were termed dehalogenases or halidohydrolases[22] which cleave the carbon-halogen bonds(s) of mono- or di-substituted compounds yielding hydrox- or oxo- carboxylic acids, respectively.

$$\underset{\text{2-chloroalkanoic acid}}{\text{R-CHCl-COOH}} + H_2O \longrightarrow \underset{\text{2-hydroxyalkanoic acid}}{\text{R-CHOH-COOH}} + H^+ + Cl^-$$

$$\underset{\text{2,2-dichloroalkanoic acid}}{\text{R-CCl}_2\text{-COOH}} + H_2O \longrightarrow \underset{\text{2-oxoalkanoic acid}}{\text{R-CO-COOH}} + H^+ + Cl^-$$

The purification of various enzymic forms from different bacteria and subsequent assays has led to the division of dehalogenases into two major classes[23].

(i) Class 1

This group of dehalogenases are further subdivided into 1D and 1L types to designate the stereospecificity for D- or L-isomeric substrates, respectively. Another characteristics of this group of enzymes is that they remove the halides from the compounds and cause an inversion in the product configuration. Most of the class 1 dehalogenases belong to type 1L and were isolated from species such as *Pseudomonas*[24,25], *Xanthobacter*[26] and *Moraxella*[27]. The type 1D dehalogenases were mainly isolated from *Pseudomonas*[28,29].

(ii) Class 2

This type of dehalogenases are able to utilize both D- and L-isomers as substrates. They are also subdivided into types with respect to the stereospecific product formation. Type 2I behaves similarly to that of class 1 dehalogenases in such a way that they invert substrate-product configuration and contain enzymes obtained from *Pseudomous* [30,31] and from *Rhizobium* [32] species. Type 2R dehalogenases were isolated from *Pseudomous*[31] and from *Alcaligenes*[33] species and differ from the 2I type on their retention on the substrate-product configuration.

Some of the class 1 and 2 dehalogenases had been cloned and sequenced and they showed quite a diversed nature without apparent sequence homology[25,27,34]. However it is still too early to draw any conclusive remark since the available data are still very limited for comparative purpose.

2.2.2 *Other Haloalkanoic Acids*

β-substituted haloalkanoic acids, e.g., 3-monochloropropionic acid (3MCPA), is not a substrate for the 2-haloalkanoic acid dehalogenases described in the earlier section. Earlier work[35] suggested that *Micrococcus denitrificans* metabolized 3MCPA to acrylic acid and not 3-hydroxypropionic acid, probably via dehydrogenation. More recent studies suggested that oxygen was required and 3MCPA was transformed to *trans*-3-chloroacrylic acid by dehydrogenase reaction and subsequent hydration with simultaneous chloride removal to yield 3-hydroxypropionic acid[36].

Other example on degradation of 3-haloalkanoic acid is on the degradation of 3-chlorocrotonic and 3-chlorobutyric acids by an *Alcaligenes* species[37].

$$CH_3\text{-}CCl\text{-}CH\text{-}COOH \xrightarrow{\text{CoA, ATP, } Mg^{2+}} CH_3\text{-}CCl\text{-}CH\text{-}COOSCoA \xrightarrow[HCl]{H_2O}$$

trans-3-chlorocrotonic acid → trans-3-chlorocrotonyl-CoA

$$CH_3\text{-}COH\text{-}CH\text{-}COOSCoA \longrightarrow CH_3\text{-}CO\text{-}CH_2\text{-}COOSCoA \rightarrow \rightarrow$$

trans-3-hydroxycrotonyl-CoA → 3-ketobutyryl-CoA

$$CH_3\text{-}CHCl\text{-}CH_2\text{-}COOH \xrightarrow{\text{CoA, ATP, } Mg^{2+}} CH_3\text{-}CHCl\text{-}CH_2\text{-}COOSCoA \xrightarrow[HCl]{H_2O}$$

3-chlorobutyric acid → 3-chlorobutyryl-CoA

$$CH_3\text{-}C\text{-}OH\text{-}CH\text{-}COOSCoA \longrightarrow CH_3\text{-}CO\text{-}CH_2\text{-}COOSCoA \rightarrow \rightarrow$$

3-hydroxybutyryl-CoA → 3-ketobutyryl-CoA

In these cases the initial reactions were not hydrolytic dehalogenation reactions and were strictly dependent on CoA, ATP and Mg^{2+}.

2.3 Degradation of Haloalcohol

Degradation of haloalcohol has been noted since 1965 on metabolism of 3-bromopropanol[38] but the study on this area is rather limited. The detail degradative mechanism still await future investigation. Today the known enzymes involved in degradation of haloalcohol were halohydrin epoxidases, haloalcohol dehalogenases[39] and haloalcohol halogen-halide lyases and halohydrin hydrogen-halide lyases [40,41].

$$CH_2Cl\text{-}CHOH\text{-}CH_2Cl \xrightarrow[\searrow HCl]{\text{haloalcohol dehalogenase}} CH_2OCH\text{-}CH_2Cl \xrightarrow{H_2O \searrow \ \text{epoxide hydrolase}}$$

1,3-dichloro-2-propanol → epi-chlorohydrin

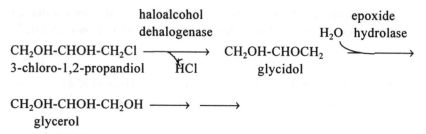

$$CH_2OH\text{-}CHOH\text{-}CH_2Cl \xrightarrow[HCl]{\text{haloalcohol dehalogenase}} CH_2OH\text{-}CHOCH_2 \xrightarrow[]{H_2O \text{ epoxide hydrolase}}$$

3-chloro-1,2-propandiol — glycidol

$$CH_2OH\text{-}CHOH\text{-}CH_2OH \longrightarrow \longrightarrow$$

glycerol

In the present there are two types of alcohol dehalogenase. Class S isolated from *Corynebacterium*[40] and *Arthrobacter*[42] species has a broad substrate specificities and the enzymes involved are simple multimeric proteins of similar subunits. The other class C, isolated from *Corynebacterium*[40] has a narrow substrate specificities and contain complex structure of at least two different subunits.

2.4 Degradation of Haloalkanes

Haloalkanes are widely distributed in our environment. Naturally occurring haloalkanes are present[43] but most of the existing haloalkanes are man-made[44]. In general straight chain haloalkanes of 10 to 18 carbons are easily degradable while liquid haloalkanes of 5 to 9 carbons and gaseous haloalkanes of 1 to 4 carbons have been studied extensively mainly because of their toxic, carcinogenic and teratogenic properties.

2.4.1 Hydrolytic Dehalogenation

Microbes involved in the degradation of haloalkanoic acids can be readily isolated from the natural environment. Bacteria capable of degradation of haloalkanes, however, are more difficult to isolate[45,46] and the hydrolytic dehalogenases can be divided into two types with respect to the substrate specificity. Class 3B enzymes had been isolated from Gram negative *Pseudomonas*[47] and from Gram positive bacteria such as *Rhodococcus*[48], *Arthrobacter*[49], *Corynebacterium*[50] species and Gram positive actinomycete[51]. These enzymes showed broad substrate specificities towards n-substituted halogenated alkanes of 2 to 10 carbons in length[52]. Class 3R enzymes were mainly isolated from Gram negative bacteria and showed rather limited substrate range. Enzyme from *Ancylobacter aquaticus*[53] and dhl-A from *Xanthobacter autotrophicus* strain GJ10[54] hydrolyze halogenated alkanes of 2 to 4 carbons only :

$$R\text{-}CH_2\text{-}X + H_2O \longrightarrow R\text{-}CH_2\text{-}OH + HX$$

In all the reported cases, the hydrolytic haloalkane dehalogenases did not dehalogenate the haloalkanoic acids, even those with the same number of carbon atoms. The reaction, as indicated, leads invariably to acidification of the medium.

For degradation of haloalkanes where the halogen position is not terminal, oxidation of the terminal methyl group to 2- or 3- haloalkanoic acids precedes the dehalogenation process[55]. The corresponding 2- or 3-haloalkanoic acids will then be degraded by the appropriate dehalogenation mechanisms mentioned earlier.

2.4.2 *Oxidative Dehalogenation*

Oxygenases are broad substrate multifunctional proteins found in both prokaryotes and eukaryotes. They are enzymes involved in a range of reactions, including dehalogenation. Many of these reactions involved the oxidation of halogenated substrate, via insertion of oxygen, to the corresponding aldehyde. Subsequent dehydrogenation forms the analogous acid[56].

$$R\text{-}CH_2\text{-}X + 1/2\ O_2 \xrightarrow{\text{Oxygenase}} R\text{-}CHOH\text{-}X \xrightarrow{\text{Spontaneous}}$$

$$R\text{-}CHO + HX \xrightarrow{\text{dehydrogenase}} R\text{-}COOH$$

This type of reaction has been found in *Pseudomonas butanovora*[55], *Corynebacterium* species[50] and in *Rhodococcus erythropolis*[57] and was found to be associated with membrane-bound system which requires electron transport for activity.

Such activities were also described for methane-monooxygenases in methane-utilizing bacteria such as *Methylosinus trichosporium*[58] and

Methylococcus capsulatus[59] and for ammoniamonooxygenases in nitrifying bacteria[60]. In general the substrates were degraded by cometabolism and therefore do not act as sole carbon and energy sources for growth[61,62].

In mammal, detoxification of halogenated hydrocarbons is also probably mediated via cytochrome P-450 monooxygenases[63]. In *Pseudomonas putida* PpG-786, a camphor induced cytochrome P-450, however, degraded bromotrichloromethane by reductive dehalogenation mechanism[64].

2.4.3 Glutathione (GSH-)-Dependent Dehalogenation

Dichloromethane (DCM) is a common industrial solvent and microorganisms isolated from soil and groundwater contaminated with this compound have been shown to contain a hexameric glutathione-dependent which convert DCM to formaldehyde and inorganic chloride[65].

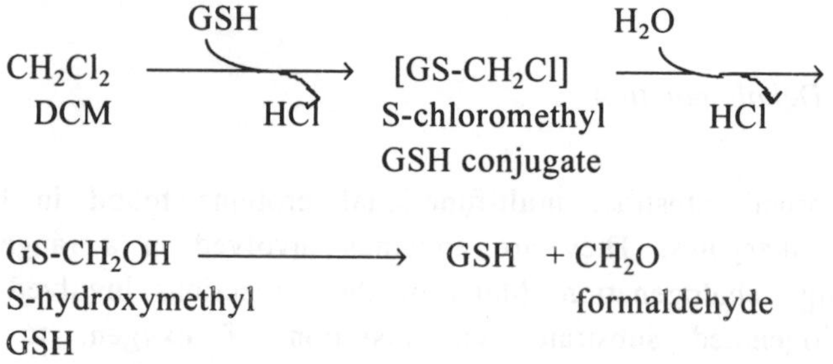

This system was analogous to the glutathione S-transferase system found in liver cytosol[66]. Moreover, amino acid sequences analysis showed three conserved regions among the glutathione S-transferases from bacteria, maize, rat and man[67] even though the eukaryotic proteins were only dimers.

2.4.4 Anaerobic Reductive Dehalogenation

In environments where the aerobic oxidative dehalogenation cannot occur micoorganisms can cleave the carbon-halogen bonds by means of reductive mechanism[68,69,70]. This reductive dehalogenation reaction is also important in the breaking down of highly halogenated compounds, which are not normally good substrates for oxidative and hydrolytic reactions. The degradation or transformation of haloalkanes such as chloroform, carbon tetrachloride, bromoethane, dibromoethane, dichloroethane, 1,1,2,2-tetrachloroethane by methanogenic bacteria such as *Methanococcus thermolithotrophicus, M. deltae* and *Methanobacterium thermoautotrophicum* has been[71] reported. A typical example is the sequential degradation of tetrachloroethylene to trichloroethylene, dichloroethylene, vinyl chloride, ethene then to carbon dioxide[72].

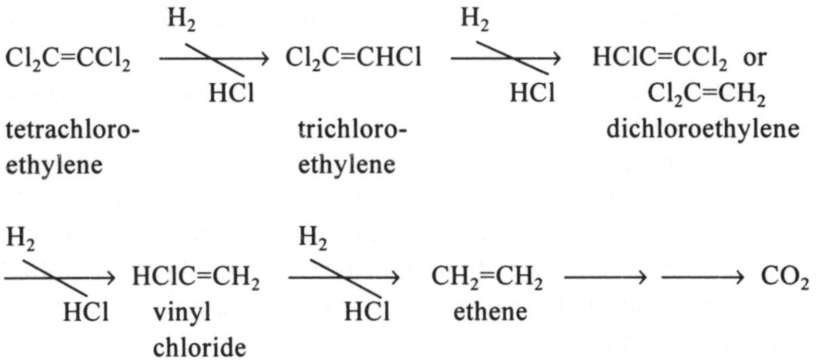

The reductive dehalogenation of tetrachloromethane by *Acetobacter woodii* and anaerobes was found to occur only in the presence of acetyl-Coenzyme A pathway and the production of acetate[73]. This, together with other information, implied the involvement of electron transfers in these reactions[74].

The reductive dehalogenation reaction can actually be divided into two types. Hydrogenolysis replaces the halogen atom with a hydrogen atom whereas compounds with halogen atoms on vicinal carbon atoms were reduced by dihaloelimination with the formation of molecules with two fewer halogens and an additional carbon-carbon double bond.

3 Biodegradation of Halogenated Aromatic Compounds

3.1 The Chemistry of Halogenated Aromatic compounds

Halogenated aromatic compounds, for exaple, polychlorinated biphyenyls (PCBs), 2,4-dichlorophenoxyacetic acid (2,4-D), 2,4,5-trichlorophenoxyacetic acid (2,4,5-T) and 1,1,1-trichloro-2,2'-bis(4-chlorobiphenyl) ethane (DDT) have been used extensively in industry and in agriculture for many years. Many of these compounds have been used as solvents, paints, varnishes, herbicides and pesticides and because of their widespread usage, have been released into the environment substantially. Some of these compounds, for example, PCBs, are strongly hydrophobic and therefore tend to accumulate in the fatty tissues causing detrimental effect while others metabolized via the liver. Others are readily degradable, sometimes slowly by various types of microorganisms. In many cases the degradation or simply transformation is mediated by means of cometabolism[76], a process where the organism cannot use the metabolized compound as a carbon or energy source but metabolized it through the use of another substrate for growth.

Halogenated aromatic compounds are more resistant to degradation because the balogen atom decreases the reactivity of the aromatic ring to electrophiles[77]. In general, however, the biodegradation of halogenated aromatic compounds is very similar to that of aromatic hydrocarbons and halogenated aliphatics. The biodegradation involves the breakage of the carbon-halogen bonds, cleavage of the aromatic ring and subsequent utilization of the carbon backbone and in many cases, through the metabolic pathways for non-halogenated aromatics.

3.2 Dehalogenation Before Ring Cleavage

The removal of the halogen atom can occur either before or after ring cleavage. In aerobic condition oxidation involves mono- or di-oxygenation whereas in anaerobic condition reduction is the major reaction. Hydrolytic degradation, on the other hand can occur in both aerobic and anaerobic conditions.

3.2.1 Oxidative Dehalogenation

This is a less common dehalogenation mechanism utilized by microbes, whereas a dioxygenase catalyze the fortuitous hydroxylation of the halogen-containing carbon and an adjacent carbon atom. The mineralization, for example, of 2-fluorobenzoate is shown below[78].

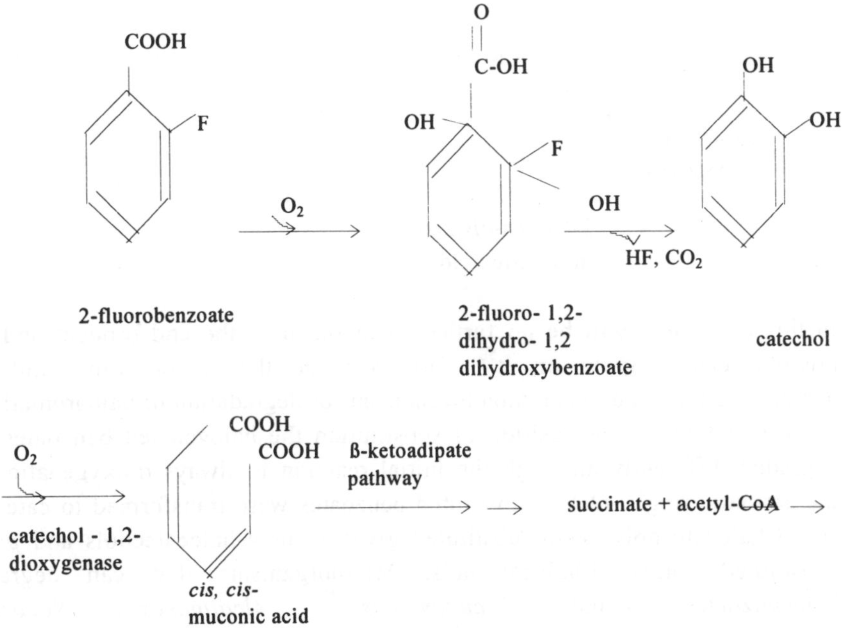

The halogenated benzoate is transformed to catechol, then *cis, cis*-muconic acid, by dioxygenases and was subsequently metabolized by the β-ketoadipate pathway for carbon and energy[79,80]. However the same compound may also be transformed to non-utilizable end-product.

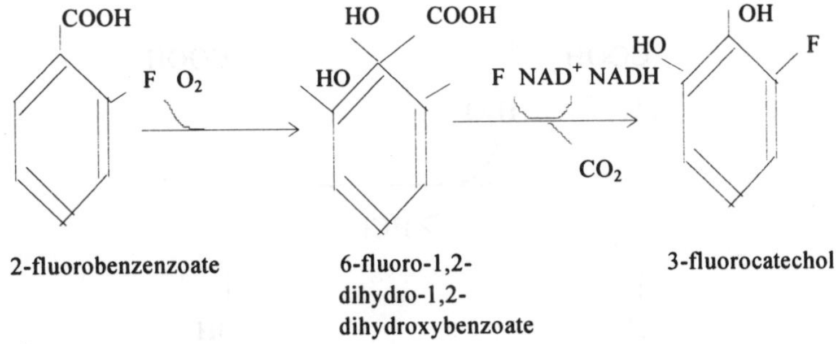

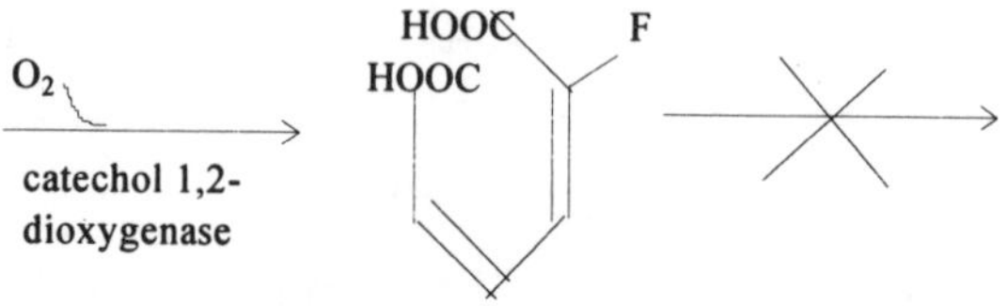

2-fluoro-*cis,cis*-
muconic acid

In this case there will be no further metabolism of the end product and the microbe cannot grow on it[56]. This may be the reason why oxidative dehalogenation is not a common mechanism for degradation of haloaromatics.

Depending on the position of substitution the halogenated benzoates are degraded differently although the initial reaction is always dioxygenation of the aromatic ring[81]. *Ortho*-substituted benzoates were transformed to catechol and 3-halocatechols, *meta*-substituted gives 3- or 4-halocatechols and *para*-substituted forms 4-halocatechols. Microorganisms that can degraded halobenzoates include *Pseudomonas*[81,82], *Alcaligenes*[81], *Nocardia*, *Flavobacterium*[83] and *Azotobacter*[84] species.

3.2.2 *Hydrolytic Dehalogenation*

Another degradative process for haloaromatics is hydrolytic dehalogenation where a halogen is replaced by a hydroxyl group[80,85].

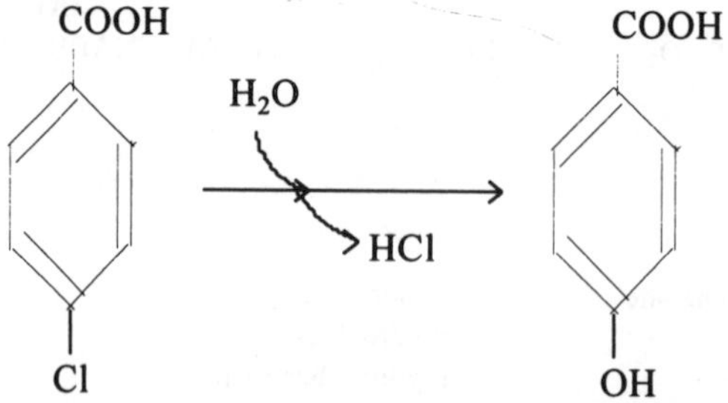

4-chlorobenzoate 4-hydroxybenzoate

For example, the conversion of 4-chlorobenzoate to 4-hydroxybenzoate by an *Arthrobacter* and a *Pseudomonas* species[79,85]. The oxygen atom in the hydroxyl group in this case is derived from water instead of oxygen.

In general, halogenated aromatics that have additional electron-withdrawing substitutions such as carboxyl (-COOH), nitro (-NO_2) and halogen (-X) *ortho-* and/or *para-* to the halogen substituent are more susceptible to degradative processes. On the other hand if the additional substitutions are electron-donating elements such as hydroxyl (-OH), amino (-NH_2) and alkyl (e.g. methyl-, ethyl-) group then the compound will be more resistant to biodegradation[56]. Microorganisms that utilized hydrolytic dehalogenation on degradation of halogenated aromatic compounds include *Arthrobacter* and *Pseudomonas*[79,85,86], *Rhodococcus*[87,88], *Mycobacterium*[88] and *Flavobacterium* species[89].

3.2.3 Reductive Dehalogenation

Reductive dehalogenation of haloaromatics is a process where the halogen atom is replaced directly with a hydrogen atom, producing a more reduced compound. This process can occur in both aerobic and anaerobic organisms.

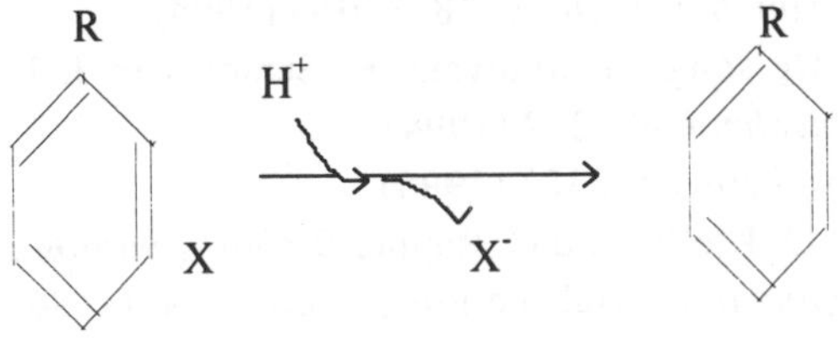

Anaerobic reductive dehalogenation was first reported in 1983[90] on the degradation of chloro-, bromo- and iodobenzoates by methanogenic communities. *Meta*-substitutions were more susceptible to dehalogenation than *ortho-* or *para*-isomers in this case. But in another case *ortho*-substituted chlorophenol was more susceptible to degradation[91]. Microorganisms that utilized reductive dehalogenation include *Desulfomonile tiedjei, Rhodococcus chlorophenolicus*[87], *Flavobacterium*[89], *Alcaligenes denitrificans*[92] or *Coryneform bacterum*[93] and *Corynebacterium pedonicum*[94].

3.4 Dehalogenation After Ring Cleavage

Some of the halogen substituents of many halogenated aromatic compounds were removed after ring cleavage. These compounds including halobenzoates, halobenzenes, haloanilines and halophenoxyacetates[56,95,96], were mainly converted to halocatechols. Mono- and di-substituted halocatechols were degraded by *ortho*-cleavage and the halides were released during subsequent metabolism[79,97].

References

1. M. Alexander, *Biodegradation and Bioremediation.* Academic Press, San Diego, (1994).
2. N. J. Bunce, *Environmental Chemistry.* Wuerz Publishing Ltd, Winnipeg. (1991).
3. G. W. Gribble, *Environ. Sci. Technol.* **28**, 310A- (1994).
4. H. E. Spinnler, E. De Jong, G. Mauvais, E. Semon and J. L. Le Quere. *Appl. Microbiol Biotechnol.* **42**, 212 (1994).
5. F. Laturnus. *Chemosphere.* **31**, 3387 (1995).
6. S. L. Neidleman, L. O. Kjeller and C. Rappe, *Biohalogenation : Principles, Basic Roles and Applications.* Halsted Press, New York. (1986).
7. G. H. Verdoorn and B.-E. van Wyk. *Phytochemistry.* **31**, 1029 (1992).
8. V. Fajardo, P. Frederico and M. Shamma. *J. Nat. Prod.* **54**, 554 (1991).
9. A. Svenson, L. O. Kjeller and C. Rappe, *Environ. Sci. Technol.* **23**, 900 (1989).
10. A. Hulster, J. F. Muller and H. Marschner, *Environ. Sci. Technol.* **28**, 1110 (1994).
11. M. J. Wang and K. C. Jones, *Environ. Sci. Technol.* **28**, 1260 (1994).
12. H. Kylin, E. Grimvall and C. Ostman, *Environ. Sci. Technol.* **28**, 1320 (1994).
13. S. Jensen, G. Eriksson, H. Kylin and W. M. J. Strachan, *Chemosphere* **24**, 229 (1992).

14. A. M. Stomp, K. H. Han, S. Wilbert and M. P. Gordon, *In Vitro Cellular & Developmental Biol. Plant.* **29P**, 227 (1993).
15. K. H. Han, S. Wilbert, M. Sakine, A. M. Stump and M. P. Gordon, *In Vitro Cellular & Development Biol. Plant.* **29A**, 42A (1993).
16. L. Pauling, *The Nature of the Chemical Bond.* 3rd ed. Cornell U. Press, Ithaca, N.y. (1960).
17. J. March, *Advanced Organic Chemistry. Reactions, Mechanisms and Structure,* 2nd ed. McGraw Hill Press, London (1977).
18. P. C. Kearney, C. L. Harris, D. D. Kaufman and T. J. Sheets. *Adv. Pest Control Res.* **6**, 1 (1965).
19. J. Trotter, in *The Chemistry of the Carbon-Halogen Bond.* ed. A. Patai. J. Wiley & Sons Ltd. N. Y. (1973).
20. A. J. Murray and J. P. Riley, *Nature.* **242**, 37 (1973).
21. J. E. Lovelock, *Nature.* **256**, 193 (1975).
22. P. Goldman and G. W. A. Milne, *J. Biol. Chem.* **241**, 5557 (1966).
23. J. H. Slater, A. T. Bull and D. J. Hardman, *Biodegradation.* **6**, 181 (1995).
24. J. S. H. Tsang, P. J. Sallis, A. T. Bull and D. J. Hardman, *Arch. Microbiol.* **150**, 441 (1988).
25. B. Schneider, R. Muller, R. Frank and F. Lingens, *J. Gen. Microbiol.* **173**, 1530 (1991).
26. J. Van der Ploeg, G. Van Hall and D. B. Janssen, *J. Bacteriol.* **173**, 7925 (1991).
27. H. Kawasaki, K. Tsuda, I. Matsushita and K. Tonomura, *J. Gen. Microbiol.* **138**, 1317 (1992).
28. J. M. Smith, K. Harrrison and J. Colby, *J. Gen. Microbiol.* **136**, 881 (1990).
29. P. T. Barth, L. Bolton and J. C. Thomson, *J. Bacteriol.* **174**, 2612 (1992).
30. K. Motosugi, N. Esaki and K. Soda. *Arch. Microbiol.* **131**, 179 (1982).
31. A. J. Weightman, A. L. Weightman and J. H. Slater, *J. Gen. Microbiol.* **128**, 1755 (1982).
32. J. A. Leigh, A. J. Skinner and R. A. Cooper, *FEMS Microbiol. Lett.* **49**, 353 (1988).
33. A. Brokamp and F. R. J. Schmidt, *Curr. Microbiol.* **22**, 299 (1991).
34. U. Murdiyatmo, W. Asmara, J. S. H. Tsang, A. J. Baines, A. T. Bull and D. J. Hardman, *Biochem. J.* **284**, 87 (1992).
35. J. M. Bollag and M. Alexander, *Soil Biol. Biochem.* **3**, 241 (1971).
36. J. E. T. Van Hylckama Vlieg and D. B. Janssen, *Biodegradation.* **2**, 25 (1991).

37. D. Kohler-Staub and H. P. E. Kohler, *J. Bacteriol.* **171**, 1428 (1989).
38. C. E. Castro and E. W. Bartnicki, *Biochim. Biophys. Acta* **100**, 384 (1965).
39. T. Suzuki, N. Kasai, R. Yamamoto and N. Minamiura, **42**, 270 (1994).
40. T. Nagasawa, T. Nakamura, F. Yu, I. Watanabe and H. Yamada, *Appl. Microbiol. Biotech.* **36**, 478 (1992).
41. J. J. Van der Waarrde, R. Kok and D. B. Janssen, *Appl. Environ. Microbiol.* **59**, 528 (1993).
42. A. J. Wijngaard, P. T. W. Van der Reuvekamp and D. B. Janssen, *J. Bacteriol.* **173**, 124 (1991).
43. P. M. Gschwend, J. K. Macfarlane and K. A. Newman, *Science.* **227**, 1033 (1985).
44. L. H. Keith and W. A. Teilliard. *Environ, Sci. Technol.* **13**, 416 (1979).
45. T. Omori and M. Alwxander, *Appl. Environ. Microbiol.* **35**, 867 (1978).
46. J. S. H. Tsang, T. Yokota, T. Omori and Y. Minoda, Ann. Report ICME 6, 101-115. Osaka U. Japan. (1983).
47. Y. Nagata, T. Nariya, R. Fukuda, K. Yano and M. Takagi, *J. Bacteriol.* **175**, 6403 (1993).
48. P. J. Sallis, S. J. Armfield, A. T. Bull and D. J. Hardman, *J. Gen. Microbiol.* **136**, 115 (1990).
49. R. Scholtz, T. Leisinger, F. Suter and A. M. Cook, *J. Bacteriol.* **169**, 5016 (1987).
50. T. Yokota, T. Omori and T. Kodama, *J. Bacteriol.* **169**, 4049 (1987).
51. D. B. Janseen, J. Gerritse, J. Brackman, C. Kalk, D. Jager and B. Witholt, *Eur. J. Biochem.* **171**, 67 (1988).
52. S. Belkin. *Biodegradation.* **3**; 299 (1992).
53. A. J. Wijngaard, K. W. H. J. Kamp, J. Van der Ploeg, F. Van der Pries, B. Kazemier and D. B. Janssen, *Appl. Environ. Microbiol.* **58**, 976 (1992).
54. D. B. Janseen, F. Pries, J. Ploeg, B. Van der, Kazemier, P. Terpstra and B. Witholt, *J. Bacteriol.* **171**, 6791 (1989).
55. T. Yokota, H. Fuse, T. Omori and Y. Minoda, *Agric. Biol. Chem.* **50**, 453 (1986).
56. B. E. Rittmann, E. Seagren, B. A. Wrenn, A. J. Valocchi, C. Ray and L. Raskin, in *Situ Bioremediation.* 2nd ed. Noyes Publ. U.S.A. (1994).
57. S. J. Armfield, P. J. Sallis, P. B. Baker, A. T. Bull and D. J. Hardman, *Biodegradation.* **6**, 237 (1995).
58. R. Oldenhuis, R. L. J. M. Vink, D. B. Janssen and B. Witholt. *Appl. Environ. Microbiol.* **55**, 2819 (1989).
59. D. I. Stirling and H. Dalton, *J. Gen. Microbiol.* **116**, 277 (1980).

60. M. E. Rasche, M. R. Hyman and D. J. Arp, *Appl. Environ. Micobiol.* **56**, 2568 (1990).
61. T. Imai, H. Takigawa, S. Nakagawa, G. J. Shen, T. Kodama and Y. Minoda, *Appl. Environ. Microbiol.* **52**, 1403 (1986).
62. C. D. Little, A. V. Palumbo, S. E. Herbes, M. E. Lidstrom, R. L. Tyndall and P. J. Gilmer, *Appl. Environ. Microbiol.* **54**, 951 (1988).
63. W. B. Jakoby, *Enzymatic Basis of Detoxification.* Academic Press, Orlando (1980).
64. T. Lam and L. Vilker, *Biotech. Bioeng.***29**, 151 (1986).
65. D. Kohler-Staub, S. Frank and T. Leisinger, *Biodegradation.* **6**, 229 (1995).
66. A. E. Ahmed and M. W. Anders, *Drug Metab Disp.* **4**, 357 (1976).
67. S. LaRoche and T. Leisinger, *J. Bacteriol.* **172**, 167 (1990).
68. J. K. Kochi, *Organometallic Mechanisms and Catalysis*, Academic Press, N.Y. (1978).
69. K. Ramanand, M. T. Balba and J. Duffy, *Appl. Environ. Micobiol.* **59**, 3266 (1993).
70. A. Neumann, H. Scholz-Muramatsu and G. Diekert, *Arch. Microbiol.* **162**, 295 (1994).
71. N. Belay and L. Daniels, *Appl. Environ. Micobiol.* **53**, 1604 (1987).
72. F. Parsons and G. B. Lage, *J. Amer. Water Works Assoc.* **77**, 52 (1985).
73. C. Egli, T. Tschan, R. Scholtz, A. M. Cook and T. Leisinger, *Appl. Environ. Microbiol.* **54**, 2819 (1988).
74. D. L. Freedman and J. M. Gossett, *Appl. Environ. Microbiol.* **55**, 2144 (1989).
75. D. N. Sonier, N. L. Duran and G. B. Smith, *Appl. Environ. Microbiol.* **60**, 4567 (1994).
76. D. Graves in *Molecular Biology and Biotechnology. A Comprehensive Desk Reference*, ed. R. A. Meyers. VCH Publisher, N.Y. (1995).
77. S. Peelen, I. M. C. M. Rietjens, M. G. Boersma and J. Vervoort, *Eur. J. Biochem.* **227**, 284 (1995).
78. P. Goldman, G. W. A. Milne and M. T. Pignataro, *Arch. Biochem. Biophys.* **118**, 178 (1967).
79. W. Reineke and h. J. Knackmuss, *Ann. Rev. Microbiol.* **42**, 263 (1988).
80. L. C. M. Commandeur and J. R. Parsons in *Biochemistry of Microbial Degradation*, ed. C. Ratledge. Kluwer Acad. Publ. Netherlands. (1994).
81. D. Berge, H. Ratnaweera and H. Efraimsen, *Water Sci.Technol.* **29**, 219 (1992).

82. K. A. Vora, C. Singh and V. V. Modi, *Curr. Microbiol.* **17**, 249 (1988).
83. L. Xun, E. Topp and C. S. Orser, *J. Bacteriol.* **174**, 5745 (1992).
84. N. Walker and D. Harris, *Soil Biol. Biochem.* **2**, 27 (1970).
85. J. D. Scholten, K-H. Chang, P. C. Babbitt, H. Charest, M. Sylvestre and D. Dunaway Mariano, *Science.* **253**, 182 (1991).
86. F. K. Higson and D. D. Focht, *Appl. Environ. Microbiol.* **56**, 1615 (1990).
87. J. H. A. Apajalahti and M. S. Salkinoja-Salonen, *J. Bacteriol.* **169**, 5125 (1987).
88. M. M. Haggblom, L. J. Nohynek and M. S. Salkinoja-Salonen, *Appl. Environ. Microbiol.* **54**, 3043 (1988).
89. J. G. Steiert and R. L. Crawford, *Biochem. Biophys. Res. Comm.* **141**, 825 (1986).
90. A. Horowitz, J. M. Suflita and J. M. Tiedje, *Appl. Environ. Microbiol.* **45**, 1459 (1983).
91. S. A. Boyd and D. R. Shelton, *Appl. Environ. Microbiol.* **47**, 272 (1984).
92. W. J. J. Van den Tweel, J. B. Kok and J. A. M. DeBont, *Appl. Environ. Microbiol.* **53**, 810 (1987).
93. P. E. J. Groenewegen, A. J. M. Driessen, W. N. Konings and J. A. M. DeBont, *J. Bacteriol.* **172**, 419 (1990).
94. G. M. Zaitsev and Y. N. Karasevich, *Mikrobiologiya.* **54**, 356 (1985).
95. D. J. Hardman, *Crit. Rev. Biotech.* **11**, 1 (1991).
96. M. A. Bhat and C. S. Vaidyanathan in *Biotransformations : Microbial Degradation of Health Risk Compounds,* ed. V. P. Singh. Elsevier Science B. V. (1995).
97. J. J. Arensdorf and D. D. Focht. *Appl. Environ. Microbiol.* **61**, 443 (1995).

TECHNOLOGICAL CONCEPTS FOR BIOREMEDIATION : AN OVERVIEW OF THE SCIENTIFIC PRINCIPLES

DATTA MADAMWAR AND MANISH GANDHI

Department of Biosciences, Sardar Patel University, Vallabh Vidyanagar-388120. Gujarat, India.

1 Introduction

In 1965 Alexander had put forward the principles of microbial infallibility which stated that no natural organic compound is totally resistant to biodegradation given the right environmental conditions. Indeed, we do not observe the accumulation of large quantities of natural organic materials in the biosphere because they are subject to microbial recycling .

The development of synthetic organic chemicals during the last several decades has, however resulted in the production of a wide variety of synthetic organic compounds that have inevitably found their way into the environment either deliberately or accidentally. Many of these synthetic organic compounds are susceptible to biodegradation due to their similarity to naturally-produced organic compounds, whereas others are less biodegradable.

Man-made compounds that are detected in the environment at unusually high concentrations are termed xenobiotics. This term is also applied to those compounds that do occur naturally, but due to man's activities, are deposited in the environment at unusually high concentrations. Such xenobiotic compounds are not readily biodegradable since their molecular structures or bond sequences are not readily recognized by existing degradative enzymes. The net result is that these compounds are either resistant to biodegradation or are metabolised incompletely or slowly and consequently persist and accumulate in the environment. Compounds that persist in the environment are termed recalcitrant (or persistent).

Societies generally become concerned with the long-term environment consequences of their actions only after industrialization has provided them with substantial economic and material affluence. Recently, it has become more widely recognized and accepted that environmental problems are not restrained by national borders. The action of one community often impinge on many others. The international conscience has, therefore, been aroused and concern about Man's use and abuse of the environment has been widely expressed.

The growth of the chemical industry in last few decades has generated a wide variety of man-made (anthropogenic) compounds. The production of specific chemicals is often associated with the formation of large amounts of undersired by-products and wastes resulting from the entropy of chemical reactions. The dangers arising from exposure and their decomposition products, and the effects and hazards they cause are unknown for the majority of commercially traded substances. Hazardous wastes pose major disposal problem for all industrialized nations and a need exists for effective and permanent solutions, particularly since millions of tons of wastes are generated each year that require some degree of detoxification before they can be safely released into the environment. Microbial degradation of such wastes is rapidly being adopted as a convenient cost-effective method for removing potential pollutants [1,2,3,4]. The efficiency of microbial processes depends on the susceptibility of the compounds in a waste effluents to microbial attack.

2 Practical Approach to the Biodegradation of Xenobiotics in the Environment

The successful removal of xenobiotic compounds from the environment by biodegradation requires a detailed understanding of the factors affecting biodegradation[1]. For example the nature of the environment containing the xenobiotic compounds to be removed is a determinative factor upon the success of xenobiotic degradation[5].

2.1 Environmental factors

(1) Xenobiotic (substrate) concentration - if the concentration is too high then toxicity problems may occur, if the concentration is too low then the rate of biodegradation may be limited by the affinity of the cells for these substrates.
(2) The presence of the necessary genetic information and hence cells with the required metabolic capabilities.
(3) The pH of the surrounding environment - since pH affects growth and metabolic rates.

(4) The temperature of the surrounding environment - since temperature affects growth and metabolic rates.

(5) Water availability - the presence of water is essential for all life forms. Water may be present but not available to the cells due to the presence of high salt or high concentration of organic molecules (e.g. sugars). This may result in the extracellular environment having a greater osmotic potential than the intracellular medium such that water would tend to leave the cell rather than to enter the cell.

(6) Availability of other nutrients required for growth - including oxygen or co-metabolites.

(7) The presence of other organic material - organic compounds that may be metabolised in preference to the xenobiotic compound or act as inhibitors or co-metabolites.

The ideal environment is one in which all conditions are constant or atleast fairly predictably constant and are, of course, amenable to the growth and metabolism of micro-organisms. This ideal situation facilitates laboratory-based studies to allow the determination of the optimal treatment regime for successful xenobiotic removal. Such a situation of constant condition rarely occurs naturally but is most likely to occur for treatment of industrial effluents at source for which the physical and environmental conditions can be controlled.

Treatment of xenobiotic compounds at municipal waste water treatment works is feasible, and usually successful, but often depends on a constant supply of the xenobiotic compounds to the system as opposed to intermittent loading of the xenobiotic due to loss of microbial capabilities since -

(1) The system continuously naturally selects and enriches micro-organisms on the basis of their metabolic capabilities according to the substrates present. The absence of a particular xenobiotic compound could result in the loss, or depletion of the micro-organisms capable of degrading it[5].

(2) Shock loading of xenobiotic compounds to the system can result in elevated concentration of the xenobiotic compound or its metabolites. They may reach toxic levels which will reduce or destroy the effectiveness of the microbial population in degrading organic compounds[6].

Waste water treatment systems may never be able to treat certain xenobiotic compounds successfully due to -

(1) The inability of the system to a suitably high enough population of cells capable of degrading the compound because the xenobiotic degraders are

unable to successfully compete for nutrients, other than the carbon source, with the normal flora present.

(2) Other waste compounds present are metabolised in preference to the degradation of the xenobiotic compound. For example the degradation of chlorophenols in the presence of phenol. Chlorophenols can be successfully degraded via the ortho-cleavage pathway. If the latter is utilised by the organism for the metabolism of any chlorophenols present, subsequent built-up to toxic levels of meta cleavage metabolites of chlorophenols, will occur [7].

The successful biodegradation of xenobiotic due to spillage on land is less predictable. This is due to the complexity of the environmental system and the constant changes occurring particularly in temperature and water availability.

Environmental problems may be tackled in two fundamentally different ways - one is to adapt processes that do not create the problem in the first place. We might call this the `front of pipe' approach. The other is to treat the problem once it has been created, sometimes referred to as 'end of pipe' approach. However, current practices of attempting to find out 'end of pipe' solutions. These xenobiotic compounds can be serious threat to the biosphere/environment due to their chemical nature or toxicity. Chemicals are thus often assessed according to their pollution potential which considers the following parameters :

- the aquatic and mammalian toxicity, carcinogenicity
- levels of production
- reactivity
- bioaccumulation (biomagnification)
- persistence or recalcitrance.

Such pollution potentials are used to assess those compounds that pose an immediate threat to the environment.

3 Treatment Strategies

Xenobiotic compounds may get released through the industrial waste into the environment. The chemical composition of industrial waste is usually complex, involving a large number of compounds. However, compared to domestic and municipal sewage, industrial pollutant load is mainly represented in less than ten key compounds[8]. Treatment systems are designed to alleviate adverse affects on the environment by removing xenobiotic compounds either by complete degradation or by converting into nontoxic form [9].

Classical biological treatment systems rely on micro-organisms able to utilise the pollutants as growth substrates, thereby removing xenobiotic compounds from the environment or from the waste. There are basically two types of biological processes: aerobic and anaerobic[10]. In aerobic systems, the organic materials are usually converted to biomass or oxidised to carbon dioxide and water. In anaerobic systems part of the organic matter is converted to biomass but the majority is converted to methane and carbon dioxide.

Bioremediation of contaminated sites is a new field of endeavor, and many new or altered technologies are appearing; nevertheless the utilization of microbial processes to destroy chemicals is neither a novel idea nor a new technology. Such processes have been used for decades for the elimination of chemicals from waste streams of industries that had biological treatment systems for their effluents and, knowingly or unknowingly, for the breakdown of chemicals from households or industries serviced by municipal waste-treatment systems. For example, activated sludge systems have been shown to be capable of removing the majority of trace contaminants from the waste water [11,12]. However, problems occur if the pollutants are present in higher concentration [13]. The ability of activated sludge systems to acclimatize in order to degrade various toxic pollutants has been known, though the mechanism of this acclimatization process has not been clear. However, it is widely believed to be the result of changes in species composition rather than enzyme adaptation or mutation; cometabolism or specific consortia probably plays an important role. The fact that many compounds were not so destroyed was not necessarily the result of the absence but rather that the systems were optimized for different purpose.

The goal of bioremediation is to degrade organic pollutants to concentrations that are either undetectable or, if detectable to concentrations below the limits established as safe or acceptable by regulatory agencies. Bioremediation is being used for the destruction of chemicals in soils, groundwater, wastewater, sludges, industrial-waste systems, and gases. The list of the compounds that may be degraded by biological agents by one or other process is very long[14,15].

4 Novel Approaches

Bioremediation process must compete with existing methods in terms of efficacy, and economics. This will form the basis of new developments and can offer an effective alternative to non-biological treatment systems in that they require only modest capital investment, have low energy input, are environmentally safe, do not themselves generate wastes, and are robust and self-sustaining.

For these reasons, where conventional biotreatment is ineffective, attention is drawn towards the development of biotreatment systems utilizing specialized bacterial systems or enzymes for the treatment of toxic or recalcitrant wastes[8,9,15,16,17,18]. For example, polycyclic aromatics (non-halogenated and benzo (a) pyrene compounds are attacked by *Cunninghamella elegans*[19,20], *Pseudomonas, Beijerinckia.*

Genetic engineering approach has also been directed to industrial waste-water treatment by creating micro-organisms that can degrade specific compounds. By genetic engineering it is possible to increase the level of specific enzyme for particular degradative process [21]. Genetic determinants for the degradation of various synthetic chemicals, for example simple chlorinated phenoxyacetic or benzoic acids, are encoded on conjugative plasmids. By exploiting appropriate combinations of degradative plasmids, strains have been constructed that effect total degradation of specific halogenated aromatics. For example, plasmid pAC 25, which permits total degradation of 3-chloro but not 4-chloro-benzoate, can cooperate with the TOL plasmid (specifying toluene and xylene degradation) for the complete metabolism of both of these chlorinated compounds[22,23]. High- copy number vector plasmid carrying active constitutive promoters of the gene of aromatic ring cleavage enzyme catechol 2-3-oxygenase have been introduced for increased expression[24]. The extended application of plasmid-assisted breeding under selective conditions in a chemostat should permit the isolation of organisms with an assortment of new degradative abilities for the utilization of particular toxic compounds. Gene cloning may be necessary for the construction of strains capable of degrading certain recalcitrant molecules that cannot be utilized by strains, developed by plasmid-assisted breeding. Furthermore, various pathways constructions generated through plasmid-assisted breeding might be obtained by cloning relevant catabolic genes. Acquisition of novel degradative functions by a wide range of soil and aquatic micro-organisms, is likely to enhance the rate of degradation of pollutants and hastens their ultimate removal from the environment. Several other techniques such as plasmid-assisted molecular breeding, selecting natural mutants from populations exposed to the waste or by artificial inducing mutation are now available. However, there are certain problems associated with the uncontrolled release of genetically modified strains.

Enzymes also have a role in the removal of single xenobiotics or removal of a particular compound from a waste water containing several chemicals. However, their life time and cost are the main bottleneck though concept of immobilization is gaining importance. Enzymes like peroxidase, cyanide hydratase are found to be important for the removal or conversion of phenol and cyanide respectively to nontoxic form[8].

There are compounds present in chemical industry wastes which cannot be removed during conventional biological effluent treatment. The search for specialised bacteria to aid treatment of industrial effluent is therefore of great importance to the chemical industry from an economic point of view. Bacteria in both pure and mixed cultures have been utilised to biodegrade various xenobiotic compounds[25]. They are often obtained by continuous enrichment[3,26]. Mixed cultures are more easily obtained than pure cultures. Some time it is more of xenobiotic compounds, where they work in a synergistic manner.

5 Technologies

There are a number of types of bioremediation treatments[27]. These include modifications of techniques used historically in conventional waste treatment as well as innovative new methods. Approaches include enhancing indigenous microbial activity by adding specially formulated fertilisers to soil or sediments contaminated with, for example, oil or other carbon-rich wastes; landfarming and composting techniques for degradation of refinery wastes and military explosive materials; the degradation of chlorinated compounds in soils by the intervention of microbial enzymes; treating PCBs in soils and sediments with micro-organisms; and degradation of recalcitrant compounds from lagoon sludge and contaminated soils in slurry reactors and other biological processing systems[4,6,10,28,29,30,31]. Groundwater and surface water contamination can be treated by encouraging the growth of indigenous micro-organisms to degrade the waste *in situ*, by the addition of oxygen or by an alternative 'pump-and-treat' technology[5,6,32]. Treatment of toxic or noxious substances (or vapours) is showing promise with bioreactors containing solid supports for micro-organisms which degrade airborne contaminants as they pass through. All of these technologies have their applications, either alone or in combination. In choosing an appropriated biotechnology for remediation of a specific hazardous waste problem, it is useful to make a selection based on reliable information.

A variety of different technologies and procedures are currently being used, and a number of new and promising approaches have been suggested or have reached advanced stages of development[5]. Some of these technologies are *in situ* treatments, in which soil is not removed from the field or groundwater is not pumped for aboveground treatment. It has an advantage of relatively low cost but the disadvantage of being less subject to control. Other bioremediation technologies require removal of the contaminated material in some manner from its original location. Such removals increase the costs modestly to appreciable, but the processes are more subject to control.

5.1 Land Treatment

Micro-organisms in soil have a broad array of catabolic activities, and a simple way of destroying pollutants is to add compounds or materials containing them to the soil and rely on the indigenious microflora[33]. This procedure, often called land farming or land treatment, and it is a procedure that has been utilised for many years. This was developed from the historic refinery practice of land farming, a method by which oily waste was spread over the soil surface to facilitate subsequent natural degradation. The considerable amounts of C is added in these wastes have the potential to support a large biomass, but the soil has too little N and P to support such large biomasses. N and P are added to the soil, often in the form of commercial fertilizers. Furthermore, the O_2 demand of the microflora increases. The overlaying air into the soil is too slow to sustain the aerobic bacteria that are mainly, responsible. The need for supplemental O_2 is satisfied by mixing the soil in some way, sometimes by simple plowing, sometimes by more thorough mixing. Arrangements are also made to provide water to maintain optimum moisture levels for aerobic organisms.

Modern soil bioremediation systems, (unlike conventional land farming techniques) do not rely on large surface areas for spreading contaminated oil and sludge[34]. Instead, solids (i.e. soil or sludge) are placed in windrows or lined treatment cells and atmospheric O_2 is supplied by tilling, forced aeration or negative-pressure systems. Inorganic nutrients (fertilizers) are applied simultaneously, either manually or by automated systems installed in the treatment cells. The advantage of such biotreatment system like this are numerous and include a significant reduction in the surface area required for treatment. It reduces the remediation time because of improved design and increased system control, and the ease of applying treatment to gas emissions if necessary. Such a system can be termed as prepared bed reactor.

The successful operation of a land treatment system is based on a programme that incorporates three phases : a feasibility study, the design and construction of facilities and the operation and maintenance of the treatment cells. Land treatment is also a means to dispose of contaminated water.

Another solid-phase treatment involves the same approach to enhancing microbial activity but relies on a different way of providing O_2. Additional air is provided by vacuum extraction of soil above the water table (the vadose zone or the unsaturated soil layer), thereby supplying the terminal electron acceptor needed by the aerobic bacteria. This process which is designed for hydrocarbon-contaminated sites, is termed *bioventing* or simply *venting*.

5.2 Composting

Composting is a biological process which depends on the optimal growth activity and interaction of a mixed population of mesophilic and thermophilic micro-organisms. Thus, the composting environment is characterised by elevated temperature (>50° C), plentiful nutrients, high moisture content (>50%), unlimited O_2 and a neutral pH. In a sense, this method combines many of the good points of incineration and land farming and minizes their disadvantages.

In composting or a treatment procedure, the polluted material is mixed together in a pile with a solid organic substance that is itself reasonably readily degraded, such as fresh straw, wood chips, wood bark or straw that had been used for livestock bedding. The pile is often supplemented with N, P and possibly other inorganic nutrients. The material is placed in a simple heap, it is formed in a long rows known as windrows, or it is introduced into a large vessel equipped with some means of aeration. Moisture must be maintained, and aeration is provided either by mechanical mixing or by some aeration device. The process is aerobic, therefore, the extent of aeration determines the rate of destruction of the waste and the level to which the temperature will rise in the composting mass[37,38]. In vessel system (a contained vessel) is desirable when compost contains hazardous chemicals. Heat released during microbial growth on the solid organic material is not adequately dissipated, and hence the temperature rises. The higher temperatures (50-60° C) are often more favourable to biodegradation than the lower temperatures that are maintained in same composts. Composting has been used as a means of treating soil contaminated with chloro-phenols. A field-scale demonstration has also shown that the concentrations of TNT, RDX and HMX in contaminated sediments placed into composts decline to a marked extent as these three explosives are biodegradable. Aerated lagoons have been employed to bioremediate refinery wastes and other materials containing petrochemicals.

5.3 In Situ Bioremediation

In situ bioremediation processes all have the common objective i.e to use microorganisms (usually aerobic) to degrade contaminants in soil and/or water with least disturbance, and are, therefore, recommended in cases where excavation or disposal of soil or groundwater is not possible or economical. This is typically achieved through manipulation of the environmental conditions on site, to stimulate indigenous catabolic microbial associations[39,40,41].

A common procedure for *in situ* bioremediation entails the introduction of nutrients and O_2 into subsurface aquifers, relying on the indigenous microflora to destroy the unwanted molecules. This process is sometimes called biorestoration. Most of the contaminated sites treated so far contain petroleum hydrocarbons as the contaminants. Leakages from underground storage tanks containing gasoline result in the appearance of benzene, toluene, ethylbenzene, and xylenes. Although these compounds are initially in the gasoline phase, particular attention is given to them because they are toxic and because they enter the aqueous phase in the form of a sustained release. This sustained release and the amounts present in the aqueous phase are consequences of their reasonable water solubilities and the constant partitioning of these compounds from the gasoline to the aqueous phases[42].

In situ treatments involving nutrients and one or another source of O_2 are utilized in many field operations. Some bacteria are able to use in addition to O_2 nitrate as an electron acceptor, and they thus grow and degrade a number of substrates in anoxic waters provided with nitrate. Nitrate is attractive because of its high solubility in water and low cost, although caution must be exercised because nitrate, if present in drinking water at levels in excess of 10 mg/L (as N), is a pollutant itself.

Many contaminated aquifers are anaerobic as a result of oxygen depletion by indigenous aerobic microorganisms, hydrocarbon degradation by indigenous anaerobic bacteria merits serious consideration at some sites as a method of groundwater restoration. In response to such environmental concerns, knowledge of the metabolic capabilities of anaerobic bacteria has expanded dramatically in the past decade, particularly with respect to single-ring aromatic hydrocarbons and closely related oxygenated compounds. For example, 18 pure cultures capable of anaerobic toluene degradation have been reported since 1989, including 16 denitrifying cultures[43], one ferric iron-reducing culture and one marine sulfate-reducing culture; ferementative-methanogenic mixed enrichment cultures that degrade toluene have also been reported[44].

5.4 Bioaugmentation

Bioaugmentation is a type of inoculation which enhances the decomposition of synthetic compounds in the environment, during bioremediation process. Fungal inocula currently used for bioremediation of contaminated soils are produced by techniques developed for the production of fungal spawn for supply to mushrooms growers or rely on "bulking up" of pure cultures of selected or agricultural wastes. These technique are labour intensive, and the quality of the inoculum produced is quite variable. There is an extensive information on different types of carriers of fungal inocula for biological control, fungal spawn, introduction of mycorrhizal fungi, and bioremediation. Examples, include peat, granular vermiculture mixtures, grains and alginate pellets[45]. Currently lignin-degrading fungi are delivered to soil via various organic substrates such as wood chips, wheat straw, corn cobs, commercial mushroom spawn, and other agricultural products that are thoroughly grown through with selected fungi. These substrates contain a nutrient reserve that supports fungal colonization of contaminated soil[46].

5.5 Aboveground Bioreactors

A variety of techniques have been described in which pollutants are treated above ground. Some are old practices, such as activated-sludge treatment of waste streams containing synthetic chemicals; others have been introduced into practice quite recently; and still others exist only in laboratory or sometimes in pilot-plant treatment units. Aboveground bioreactors are used as a means of remediation where remedy is sought only after a problem is found, and many of these reactors are designed to destroy chemicals in industrial waste streams before those compounds reach natural environments.

Bioreactor technology is broad in applications, competitive and, still innovative. Researchers are constantly striving to achieve the design of the absolute bioreactor: the reactor capable of fitting every possible clean-up situation. The bioreactor sector of the bioremediation industry is one in which engineers and scientists can create a near-perfect environment for biodegradation, through manipulation of the elements of nature. The close ties of this approach to conventional favorable perception by both environmentalists and engineers.

Simplistically, a bioreactor is a reaction vessel which has a system for delivering oxygen and nutrients and devices for thorough mixing and the adjustment/maintenance of pH. Also, the reactor is normally fitted with influent and effluent pumps and can be run in a batch or continuous mode.

The types of bioreactors used for bioremediation are numerous and include submerged fixed-film, plug flow, fluidised bed, biofilters, biological rotating contact reactors, multichamber and sequencing batch reactors.

The principal advantages are reduced sludge production and a more rapid degradation rate, probably due to the greater surface area of the biofilm. However, caution must be exercised while using these techniques as non-biodegraded compounds may be sorbed onto the biofilm and the support matrix making the generated sludge a potential hazardous material. Bioreactors are found to be most suitable to genetically engineered or specialized microorganisms because of the enclosed design that facilitates greater control of the organisms.

5.6 *Slurry Bioreactors*

One reactor which has particular significance in the bioremediation is the soil slurry reactors. These reactors are expensive and the soil needs to be pretreated before entry into the vessel. However, they show greater promise.

Silts and clays or sludges, are most suitable for this type of treatment as the soil must be smaller than 60 mesh so that an adequate slurry can be maintained against gravity. As treatment is completed during several days retention time, the soils must be dewatered by centrifugation or a belt-filter press.

In slurry-phase treatment contaminated solids are mixed constantly with a liquid and shows effective bioremediation. The system may be reasonably unsophisticated and entail introduction of the contaminated soil, sludge, or sediment into a lagoon that has been constructed with a liner, or it may be a sophisticated reactor in which contaminated materials are mixed[47]. It has close resemblance to the activated sludge procedure, and it allows the aeration, adequate mixing and control of many of the factors affecting biodegradation. Some designs allow for the capture of volatile organic products that may be generated. The level of dissolved O_2, the pH and the concentration of inorganic nutrients may be monitored and controlled. Some bioreactors are inoculated with a single species or a mixture of microorganisms able to function effectively under the controlled conditions. System has been successfully tried with PAHs, heterocycles, and phenols[25].

Since bioreactors operate at near-ideal conditions for biodegradation, contaminant removal proceeds at a rapid rate and, therefore, offers significant advantages compared with conventional techniques such as land farming[48].

5.7 *Fixed films, Immobilized Cells, Plug-Flow Reactors*

These reactors have the reputation of being highly effective and adaptable. As a result, they can be successfully used to treat low to high concentration of organics found in groundwater and in treatment of wastewater where the organic loading can exceed 1000 mg/L[49]. Their success lies in their design flexibility. It can withstand extreme fluctuations in organic loading which maintains an active biomass for long periods of time[50].

A common way of bringing about biodegradation is to use reactors in which microbial cells become attached as a film to some matrix. A solution containing the chemicals is passed over the resulting biofilm, which brings about a rapid biodegradation because of the high cell density.

A modification of fixed-film treatment employs immobilized or strongly sorbed cells. The cells are immobilized by firmly attaching the organisms or physically embedding them in the solid matrix. The cells may thus be immobilized in or on a variety of materials, including alginate beads, diatomaceous earth, hollow glass, glass fibers, membrane hollow tubes, polyurethane foam, activated C and polyacrylamide beads.

Common to many of these systems is the greater tolerance to high chemical concentrations of the cells that are in the films or that are immobilized than cells in suspension. The greater resistance may be associated with sorption of the substrate to the solid or immobilizing material, thereby reducing the amount available to suppress the microorganisms, or to some other mechanism. A number of compounds are readily biodegraded by these procedures[51].

5.8 Biofilters (Vapor Phase Bioreactors)

Microorganisms are also being used to destroy a variety of volatile compounds. In such technologies, the microorganisms are allowed to grow on some solid support and stream of gas containing the unwanted molecules is passed through the solid support. The resulting microbial action leads to destruction of the contaminants. In Europe, biofilteration has become an accepted technology for treating volatile organic compounds and odor-containing industrial exhaust[52]. It is now a proven technology that is very economical for high volume emissions with low concentrations of pollutants. It is especially attractive because of its low energy consumption and low maintenance requirements.

A large number of volatile compounds can be degraded in biofilters, including naphthalene, acetone, propionaldehyde, volatile S compounds, toluene, benzene, dichloromethane, and vinyl chloride. Particular attention has been given to volatile and offensive odors. Biofilters or bioscrubbers can also be operated to destroy volatile chlorinated compounds.

There are two types of biofilters: the soil filter and the treatment bed/disk. The former is the simple in design. Contaminated air is passed through a nutrient-supplemented compost pile which facilitates catabolism by indigenious mesophilic bacteria. In the treatment bed, the waste, air stream and filter are humidified as gas is passed through one, two, or more beds made up of compost, wood chips, refuse, sand or diatomaceous earth. In the disks containing activated charcoal, nutrients, peat, microorganisms and compost, is placed inside a reactor vessel. Biodegradation of the contaminants occurs as the gas is vented through the system.

A carefully, engineered biofilter must be enclosed, insulated and have together with monitors for temperature and pressure of the waste streams. The filter medium must contain the essential nutrients, carbon, nitrogen and fibre to be properly operational.

5.9 *Anaerobic Processes*

Nearly all bioremediations in practical use are aerobic. However, it is now clear that anaerobic bacteria are able to catalyze many reactions and destroy many compounds that are resistant to aerobes. Particular attention in recent years has been given to chlorinated molecules, because of not only their toxicity but also their persistence in polluted environments. Anaerobes can reductively dehalogenate chlorinated molecules that persist and that are rarely attacked by aerobic bacteria[53]. Highly chlorinated compounds which are ubiquitous to less chlorinated compounds. A two-stage anaerobic-aerobic biofilm reactor proposed for treating groundwater and industrial effluents containing highly chlorinated molecules. An aerobic process has also been suggested for the on-site treatment of soil containing toxaphene, but the process, which involves addition of large amounts of organic matter to create anaerobic conditions, is slow and incomplete.

Through investigation is needed to make anaerobic bacteria into practical methods for bioremediation.

It is very important to understand that there is unlikely to be a single technology that will work in all situations or that will by itself totally remediate a complex waste, rather needs a sequence of applications of different technologies.

Microbiological clean-up is a developing technology founded upon basic principles of microbial ecology and physiology. The ability of microorganisms to reduce the potential toxicity of substances to higher organisms through processes of biodegradation, biotransformation and bioaccumulation have been successfully demonstrated in the laboratory.

In theory, there is no reason why biotechnologies based on such capabilites cannot be successfully developed and applied. However, these processes may be limited by environmental conditions. Because of increased public awareness and pressure from legislators to control the release of industrial effluents into the environment, biotechnological methods for bioremediation will undoubtedly play an increasing important role in the future.

References

1 M. Alexander, *Science.* **211**, 132 (1981).

2 C.P. Leslie Grady, Jr. *Biotech. Bioengg.* **27**, 660 (1985).

3 D. Liu, R.J. Maguise, and G.J. Pacepavicius, *Environmental Toxicology and Water Quality: An International Journal.* **7**, 355 (1992).

4 D.K. Sharma, *J. Scientific & Industrial Research.* **54**, 582 (1995).

5 Z.M. Lees, and B. Senior, in *Clean Technology and the Environment* eds R.C. Kirwood and A.J. Longly, Chapman & Hall, U.K. (1995).

6 M. Alexander, *Biodegradation and Bioremediation*, Academic Press, Inc., California, USA. (1994).

7 Biotechnological Innovations in Energy and Environmental Management, Butterworth Heinermann, Oxford, U.K., (1994).

8 J.M. Wyatt, *Microbiological Sciences.* **5**, 186 (1988).

9 J.S. Knapp, and P.S. Newby, *Wat. Res.* **29**, 1807 (1995).

10 R.E. Speece, *Environ. Sci. Technol.* **17**, 416A (1983).

11 M.J. Stewart, *International Pollution Control Magazine.* **1**, 90 (1972).

12 P. Vochten, D. Schowanck, and W. Verstraete, in *Anaerobic Digestion* eds E.R. Hall and P.M. Hobson, IAWPRC, London, (1988).

13 P.M. Bethouex, and R. Fan, *J. Water Pollu. Control Fed.* **58**, 368 (1986).

14 Z. Liu, A.M. Jacobson, and R.G. Luthy, *Appli. Environ. Microbiol.* **61**, 141 (1995).

15 P. Kumaran, *Journal of Indian Association of Environmental Management.* **20**, 15 (1993)

16 K. Kobayashi, and Rittman, *Wat. Res.* **24**, 75 (1982).

17 G.A. Lewandowshi, P.M. Armenante, and D. Pak, *Wat. Res.* **24**, 75 (1990).

18 K. Furukawa, and N. Kimura, *Environmental Health Perspective.* **103**, 21 (1995).

19 S. Dagley, in *Degradation of Synthetic Organic Molecules in the Biosphere* National Academy of Sciences; Washington (1972).

20 Natural Resources Defense Council **8**, ERC, 2120, 2122-2129, D.D.C (1975).

21 K. Furukawa, S. Hayashida, and K. Taira, *Gene.* **98**, 21 (1991).
22 W. Reineke, and H.J. Knackmuss, *Nature.* **277**, 385 (1979).
23 D.K. Chatterjee, S.T. Kellogg, D.R. Watkins, and A.M. Chakrabarty, in *Molecular Biology, Pathogenicity and Ecology of Plasmids* eds S.B. Leavy, R.C. Clowes and E.L. Koening, Plenum Press, New York (1981).
24 J.S. Karns, J.J. Kilbane, D.K. Chatterjee, and A.M. Chakrabarty, in *Genetic Control of Environmental Pollutants, Basic Life Science,* eds G.S. Omenn and A. Hollaender, Plenum Press, New York (1984).
25 J.G. Mueller, S.E. Lantz, B.O. Blattmann, and P.J. Chapman, *Environ. Sci. Technol.* **25**, 1055 (1991).
26 C. Brilon, W. Beckman, M. Hellwig, and H.J. Knackmuss, *Appl. Environ. Microbiol.* **42**, 39 (1981).
27 Al. W. Bourquin d`apre`s, Bioremediation of hazardous waste, Biofuture. **24**, (1990).
28 J.K. Mihelcic, and R.G. Luthy, *Appl. Environ. Microbiol.* **54**, 1182 (1988).
29 G.K. Anderson, and M.B. Percod, in *The Treatment and Handling of Wastes*, eds. A.D. Bradshaw, Sir R. Southwood and Sir F. Warner, Chapman & Hall, London. (1992).
30 W. Verstraete, and B. Top, in *Holistic environmental Biotechnology in Microbial Control of Pollution* eds J.C. Fry, G.M. Gadd, R.A. Herbert, C.W. Jones and I.A. Watson-Craik Cambridge University Press (1992).
31 A.T. Bull, in *The Treatement and Handling of wastes*, eds A.D. Bradshawm, Sir R. Southwood and Sir F. Warner, Chapman & Hall, London (1992).
32 R. Bewley, *Chem. Ind.* **24**, 354 (1990).
33 G.B. Wilson, J.F. Parr, J.M. Taylor, and L.J. Sikora, *Biocycle.***37** (1982).
34 W.W. Hildebrandt, and S.B. Wilson, in: Proceddings of 1990 SPE California Regional Meeting, Ventura (1991).
35 C.W. English, and R.C. Loehr, *J. Hazardous Materials.* **28**, 55 (1991).
36 R.E. Hinchee, D.C. Downey, R.R. DuPont, P. Aggerwal, and R.N. Miller, *J. Hazardous Materials.* **27**, 315 (1994).
37 S.L. Crawford, G.E. Johnson, and F.E. Goetz, in: Proceedings of the Second International On Site and In Situ Bioreclamation Symposium, San Diego, CA (1993).
38 R.T. Williams, and K.R. Keehan, in *Science and Engineering of Composting: Design, Environmental, Microbiological and Utilization Aspects* eds Hoiling and Keener, Renaissance Publications, Worthington, OH, pp. 363 (1993).
39 A.K. Kaufmann, HAZ-NEWS. Jun-July, Hazardous Waste Association of California (1986).
40 M.D. Lee, J.M. Thomas, R.C. Borden, P.B. Bedient, C.H. Ward, and J.T. Wilson, *CRC Critical Review of Environmental Control.* **18**, 29 (1988).
41 B. Zacharias, E. Lang, and H.H. Hanert, *Water Research.* **29**, 1663 (1995).

42 R.B. King, G.M. Long, and J.K. Sheldon, *Practical Environmental Biormemdiation*, Lewis Publishers, CRC Press Inc., Boca Raton, Fl (1992).
43 R. Rabus, and F. Widdel, *Arch. Microbiol.* **163**, 96 (1995).
44 H.R. Beller, A.M. Sparmann, P.K. Sharma, J.R. Cole, and M. Rehinhold, *Appl. Environ. Microbiol.* **62**, 1188 (1996).
45 V. Sasek, O. Volfova, P. Erbonova, B.R.M. Vyas, and M. Matucha, *Biotechnol. Lett.* **15**, 521 (1993).
46 D. Lestan, and R.T. Lamar, *Appl. Environ. Microbiol.* **62** 2045 (1996).
47 D. Ross, *Remediation.* **1**, 61 (1991).
48 W.J. Catallo, and R.J. Protier, *Water Science and Technology.* **25**, 229 (1992).
49 M.F. Hamoda, and A.A. Al-Haddod, *Water Science and Technology.* **20**, 131 (1989).
50 G.A. Lewandowshi, *J. Water Poll. Control Fed.* **62**, 803 (1990).
51 E.J. Bouwer, and P.L. McCarty, *Biotechnology and Bioengineering.* **27**, 1564 (1985).
52 M. Duncan, H.L. Bohn, and M. Burr, *J. Air Pollution Control Association.* **32**, 1175 (1982).
53 K. Ramanand, M.T.M. Balba, and J. Duffy, in: Proceedings of the second On Site and In Situ Bioreclamation Symposium, San Diego, CA (1993).

MICROBIAL TECHNOLOGY IN PROTECTION AND REMEDIATION OF ECOSYSTEM

JITENDRA D DESAI AND HEENA DAVE

Environmental Science and Applied Biology Division,
Research Centre, Indian Petrochemicals Corporation Limited,
Baroda - 391 346 (India)

1 Introduction

We live in Chemical era, with the production of synthetic chemicals doubling every 8 years from last 4 decade. World chemical process industry (CPI) sale has touched $ 192 billion and by 2000 it is expected to be $ 2.0 trillion. In developing countries like India, CPI represents 12.5% of total industrial production and has a major impact on the national economy. Rapid industrialization in developing countries especially in Asia is of a major concern as this region possesses the most valuable flora and fauna of the world and about half the human population of the world. Moreover, PI is predicted to grow at much faster rate in this region as it has per capita consumption of chemicals 1/10th of the developed nations. Chemical manufacturing processes inevitably lead to generation of some kind of waste and waste becomes pollution when it exceeds the carrying capacity of the environment. Industries use water, land and air for their manufacturing activities and most of these resources are returned to the environment, but often heavily contaminated with a complex set of oxygen demanding materials. These substances have serious ill effects on the ecosystems in general and on the human health and environment in particular. As a result most Governments have enacted laws to regulate the entry of such chemicals by stipulating standards for the discharge of solids, liquids and gases in the environment.

Many technological options are available to meet the challenge in maintaining the ecological balance. Essentially, they can be classified as either separation or destructive/transformation processes. Separation processes concentrate the pollutant into one stream, leaving a lower residual concentration in the original waste. These processes are often employed as the pretreatment step. They include filtration, phase separation, precipitation, air and steam stripping, adsorption on activated carbon, ion exchange, membrane separations etc. In general, due to the economic and ecological reasons separation processes are employed where recovery and recycling of materials are possible. In destructive/transformation processes the harmful chemicals are changed chemically and degraded to smaller molecules. These processes are useful when the recovery/separation of pollutant is difficult. This includes thermal oxidation, incineration, chemical oxidation and biological oxidation.

Among the destructive processes, biotechnological processes play very important role in prevention of damage and restoration of damaged ecosystems, as they result in complete mineralization of harmful pollutants in cost effective way. The latest report of Organization for Economic Co-operation and Development (OECD) on biotechnology for a clean environment estimates world market to double in next five years from $ 60 billion to $ 125 billion. Currently, major research and technological activities are aimed to either prevent the entry of such harmful chemicals to the minimal level or on the remediation of the ecosystem once these chemicals have entered the environment. In this chapter an attempt has been made to discuss briefly the microbial technologies available to solve some of the above issues.

2 Microbial Technological Options

Biotreatment coupled with physico-chemical treatment is the most cost effective and promising waste management technology in reducing the toxic load on the ecosystems. Several tests are available to determine whether the waste in question is biodegradable or not. The simplest way is to compare Chemical Oxygen Demand (COD) with Biological Oxygen Demand (BOD). Generally, BOD/COD value over 0.6 indicates that the waste can be easily treated microbiologically; 0.3 to 0.6, indicates degradability by adapted bacteria and below 0.3, indicates that such a waste is practically not amenable for biotreatment.

Among microorganisms, bacteria play a key role in biodegradation of hazardous compounds. They attack hazardous organic compounds in one of the following ways;

a) Mineralization In mineralization a compound is directly converted to harmless organic molecules such as carbon dioxide, water and biomass.

b) Mineralization with Co-metabolism in this case mineralization by microorganisms takes place in presence of some other organic compounds which are essential for growth or induction of enzymes necessary to degrade the target pollutant.

c) Bio-transformation here the conversion of pollutant takes place but the resultant compound may be toxic or recalcitrant to further degradation.

For the successful commercialization of the microbiological technology a consortium of microorganisms is generally employed to achieve complete mineralization of the pollutants. The available microbial technological options for maintaining ecological balance can be grouped into following categories :

1) Waste water biotreatment technologies
2) Bioremediation technologies
3) Biofiltration technologies.

2.1 Waste water biotreatment technologies

Activated sludge process was developed in Manchester towards the beginning of this century for sanitary waste treatment using natural biodegradation of pollutants by mocroorganisms. The basic process consists of an aeration basin where wastewater and active microorganisms (sludge) come in contact and a settler where biological flocs are separated. The thickened sludge is partly recycled into the aerator to sustain active biomass. This process perhaps, could be described as the last major technology for wastewater treatment, since subsequent successful changes like; step aeration, tapered aeration, contact stabilization, oxidation ditch, extended aeration, etc. have largely been variations of the original concept. However, the work on attached growth system has recently provided the foundation for engineering developments and generation of robust wastewater treatment technologies. Biotreatment processes differ from one another in many respects, such as nature of electron acceptor (aerobic, oxygen as electron acceptor or anaerobic, nitrate, sulfate, iron etc. as electron acceptor), and biomass state (suspended or attached growth) and hydraulic regime (plug flow or completely mixed). The principal biotreatment technologies are listed in Table-1. Some of the advanced biotechnological processes for industrial wastewater treatments are briefly discussed in the subsequent section.

Tabel 1 : Biological wastewater treatment processes

Type	Aerobic	Anaerobic
Suspended growth	Activated Sludge, Oxidation Ditch, Aerated Lagoon	Anaerobic contact reactors Anaerobic pond
Attached growth	Trickling filters Plastic media biofilters Rotating biological contactors	Anaerobic upflow/downflow fixed film reactors Anaerobic RBC Upflow anaerobic sludge blanket
Hybrid	Fluidized bed reactors	Anaerobic fludized bed reactors

2.1.1 Rotating biological contactors

The most commonly used aerobic attached growth system in wastewater biotreatment is the trickling filter, but for the industrial effluent treatment rotating biological contactor (RBC) has gained more attention in recent years due to efficient COD removal. In this system, contactor consisting of several discs mounted on a horizontal shaft, partially submerged in effluent is rotated at a slow speed (Figure 1a). Microbial film is developed on the discs which alternately come in contact with air and waste water as contactor rotates. The system has short hydraulic retention time, simple process control and low energy requirement. The use of polyurethane foam (PUF) as a porous biomass support has been reported as the emerging technology by US-Environmental protection agency (EPA) and RBC coupled with PUF has been found to be most suitable for the removal of phenols and ammonium nitrogen from industrial effluents.

2.1.2 Plastic media bio-tower

Plastic media bio-tower (Figure 1b) is an advancement over the conventional trickling filters for aerobic biological treatment to remove BOD, phenols and dissolved organics. In this case stone is replaced by plastic media which provide more surface area for bio-film to grow and more voidage for the ventilation. Bio-tower is a tall (15 meters high) structure and the effluent treatment plant of higher capacity in a smaller space can be constructed. This technology is very useful for the refinery effluent treatment with about 80-90% efficiency at very low energy cost.

2.1.3 Fluidized bed process

A hybridization between the activated sludge and bio-tower led to the development of the fluidized bed process in which bacteria are grown in fluidized bed containing sand or other particulate media like granulated active carbon, wooden chips, plastic torroids, pads, cubes etc. The system consists of a tall tower having a height of 10-20 meters, thus, requiring a smaller area for the treatment plant (Figure 1c). Fluidized particles provide large surface area/ unit volume for microbial growth. The system has a very high biodegradation efficiency as the solubility of oxygen is very good. The process is now being commercially used for the treatment of wastewater from food, sugar and pharmaceutical industries.

2.1.4 Deep shaft process

The process is designed to hasten the decomposition of organic matter from effluent by increasing oxygen supplying capacity (oxygen transfer rate upto 3 kg./m^3/h) by a high hydrostatic pressure. The aeration part consists of a vertical shaft about 6 meters in diameter and 50-150 meters in depth having riser and downcomer sub-sections (Figure 1d). Compressed air is injected to provide oxygen and the circulation rate is adjusted in such a way that a longer residence and contact time is achieved. The energy and land requirements for deep shaft process is about 30-50% lower as compared to the conventional activated sludge process.

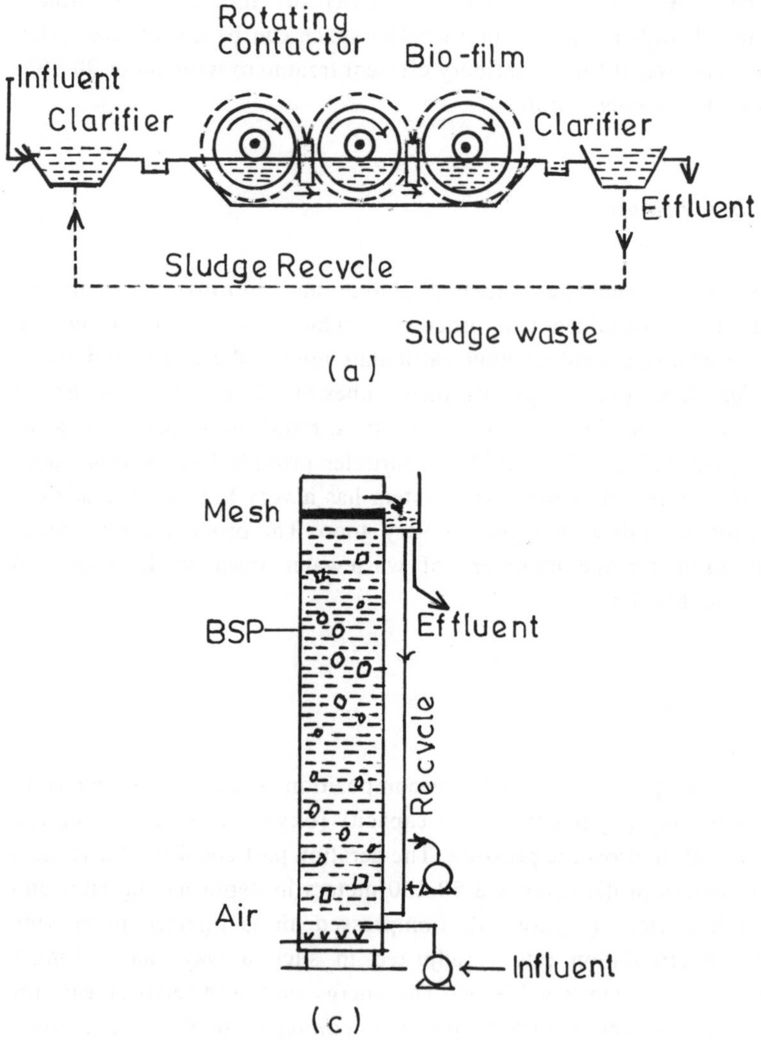

Figure 1 : Schematic diagram of advanced biotechnological processes for industrial waste water treatment (a) Rotating biological contactor process (c) fluidized bed process.

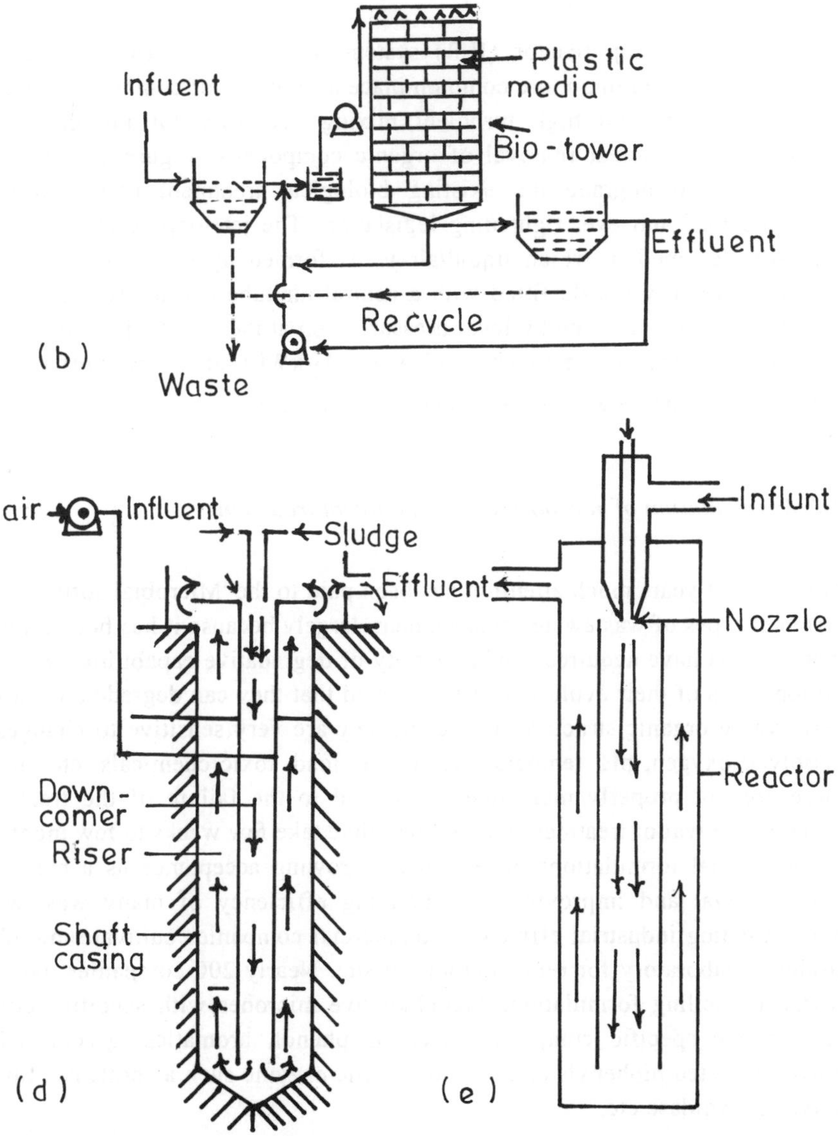

Figure 1 : Schematic diagram of advanced biotechnological processes for industrial waste water treatment (b) Bio-tower process (d) Deep shaft process and (e) Jet reactor process.

2.1.5 Jet loop reactor

The use of jet loop reactor (JLR) system for the biological treatment of wastewater is becoming more common place as a means of combining efficient oxygen transfer with high turbulent mixing necessary for the growth of microbes and efficient removal of organic compounds (Figure 1e). JLR are very useful to upgrade the existing biological treatment plants to meet increased load and ever tightening legislation. The jet loop reactor vessel is specially designed in which liquid/air jet is formed by pumping mix liquor through a ventury nozzle fitted with a central air tube and air is drawn in to liquid jet through air tube by local pressure drop at the nozzle tip. Dip to their high exygen transfer efficiency (3-4 kg.O_2/kWh) JLR are preferred to conventional surface aerators and fine bubble aerators.

2.2 Augmentation of microorganism in effluent treatment

In the recent years much attention has been paid to the 'Microbial software' for different types of wastewater management, largely because it has been realized that bacteria have acquired a wide variety of degradative capabilities during 4 billion years of their evolution. It is assumed that they can degrade almost any perceivable organic structure. However, they are very sensitive to changes in dissolved oxygen, pH, temperature, organic load, toxic chemicals, etc. and if these are not properly maintained may lead to the failure of the treatment process. Revival of treatment plants some time take few weeks to few months.

Microbial formulations have found a growing acceptance as a means of quick revival and improving the operating efficiency of many wastewater plants treating industrial effluents. Commercial companies cultivate microbial strains in laboratory for reinoculation on site. Nearly 200 companies over the world are selling formulations including live microbes with specific activity against one specific component such as phenol, aromatics, glycol, DDT, polychlorinated biphenyls etc. or one specific enzyme such as protease, lipase, amylase, cynidase etc.

3 Bioremediation Technologies

Bioremediation, the use of microorganisms to degrade organic pollutants is emerging as one of a several fastest growing technologies in removing environmental pollutants, restoring contaminated sites and preventing further pollution. Bioremediation techniques exploit the abilities of microbes to oxidize various organic compounds to reduce the toxicity, mobility or volume of the contaminant rather than merely provide temporary protection via isolation and containment. The typical end products of the process are carbon dioxide, water and bacterial biomass. Bioremediation is an attractive alternative to physical removal and subsequent destruction by incineration which is at least ten times costlier to biotreatment. Bioremediation can be applied to many types of contaminated media, like groundwater, sludges and soils. However, bioremediation of contaminated soils and groundwater offer a cost effective answer.

3.1 Stages of technology development

Essentially there are three major stages for the successful bioremediation technology development which include; site characterization, treatability and commercialization.

a) Site characterization During this stage the assessment is made, whether bioremediation technology is appropriate or not for restoration of damaged ecosystem by analyzing the chemical properties of the contaminant, and physico-chemical and biological properties of damaged site.

b) Treatability study Treatability study draws a design criteria based on investigation on kinetics of bioremediation for achieving desired results and the cost for full scale operation.

c) Commercialization stage In this stage the treatment approach is planned, equipment specifications are drawn, procurement, fabrication and pre-commissioning activities are completed and guarantee run for different operations is performed.

3.2 Technologial approaches

Bioremediation of contaminated soils and ground water can be done either *ex-situ* by excavating the ecologically damaged soil/pumping-out the contaminated ground water from aquifer followed by their treatment or by *in-situ* treatments. Basically there are two approaches namely; 1) the microbiological approach and 2) the ecological approach.

3.2.1 Microbiological approach

This approach involves augmentation of contaminated site with one or more species of contaminant degrading microorganisms. There are two methods to achieve bioaugmentation. The first involves use of pre-packed, contaminant specific degraders. Number of companies market pre-packed microorganisms with the capabilities to degrade a variety of organic contaminants. Second method involves selection, culture and application of site-specific strains that exhibit desired degradative potential for the pollutant in question.

3.2.2 Ecological approach

In this case identification and adjustment of the physical and chemical factors that are impeding the rate of degradation of the contaminant(s) by the naturally occurring microbes in the affected site is carried out. The importanat rate limiting factors of the process include;
Microbiological profile, Temperature, Oxygen, Nutrients, Bioavailability of pollutants, Chemical nature of the pollutanat, Soil population ecology.

3.3 Types of bioremediation

The available bioremediation technologies can be classified based on the operational feasibility into two types namely; *in-situ* bioremediation and *ex-situ* bioremediation.

3.3.1 In-situ bioremediation

In-situ bioremediation has been used in the petroleum industry for nearly twenty years and the basic approach has remained unchanged. Most *in-situ* bioremediations use naturally occurring organisms of the contaminated site which are adapted readily to change in their environment caused by the pollutant. Nutrient formulations containing phosphate, reduced nitrogen and trace elements and oxygen as sparged air, oxygenated water or hydrogen peroxide is introduced in the aquifer via injection well and circulated through the contaminated zone by pumping through one or more extraction wells. Bioremediation of soil and ground water is carried out simultaneously or by recirculating the groundwater using a system of injection/percolation and extraction wells as depicted in Figure 2. Optimum condition for biological activity include pH values between 6.5 and 8.5, temprature between 27° C and 35° and a ratio of organic carbon to available nitrogon and phosphorus of 300:15:1.

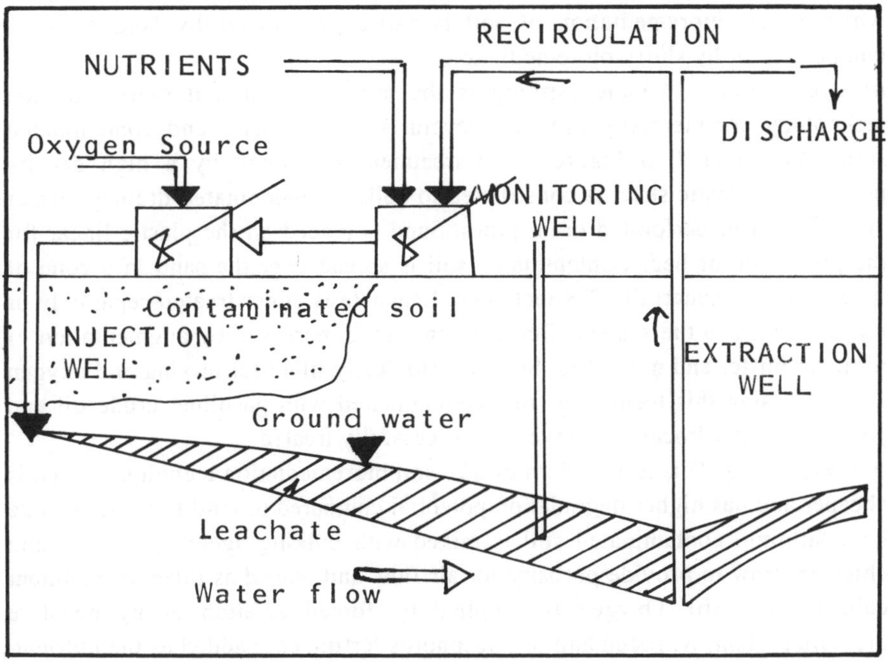

Figure 2 : *In-situ* bioremediation.

Supply of oxygen into the aquifer presents many technological difficulties and is expensive. Anaerobic processes are therefore considered as an attractive alternative to the aerobic process. Under anaerobic condition, most organic compounds are degraded by a group of interacting microorganisms, each performing a specialized reaction resulting in their complete degradation. During anaerobic degradation,nitrate, sulfate and/or Fe III can act as electron accepctors and thereby allow microbial oxidation of contaminats. While aerobic treatments of aquifers contaminated with hydrocarbons have been carried out in a number of cases, anaerobic biodegradation processes are yet in the development stage.

3.3.2. Ex-situ bioremediation

In this type of bioremediation, contaminated soil or ground water is removed and treated on-site. Contaminated water is pumped-out and treated in the same manner as for the industrial effluents described in section 2 of this chapter, while *ex-situ* bioremediation of soil is carried out either by land farming, composting or by slurry bioremediation.

a) Land farming Land farming is the most conventional bioremediation technology commercially used for several kind of wastes and contaminated soils. As depicted in Figure 3a, a treatment bed lined by a high density polyethlene plastic sheet is constructed to collect contaminated drainage water and direct it to perforated drain pipes. Sand is placed on the plastic linear for the protection of bed. Contaminated soil is spread over the sand in a zone of incorporation, generrally 2-3 feet deep. Leachate is either treated separately or recycled back to the system. Degradation rate is increased by augmentation of nuritent, buffer and mocrobes. Soil is periodically tilled for mixing and oxygen contact. Using this technique soils contaminated with gasoline, crude oil and polyromatic hyrdrocarbons have been sucessfully treated.

b) Composting This is a modern *ex-situ* soil bioremediation technique which is cheaper and has higher degradation potential compared to land farming. Unlike land farming, contaminated soil is mixed with bulking agents such as wood chips or straw to provide porosity for air flow and placed as piles or treatment cells (Figure 3b). Oxygen is supplied by forced aeration or by negative pressure system. Nitrogen and/or phosphorus fertillizer is added as the nutrient. Although agitation and oxygen contact is poor in this system,

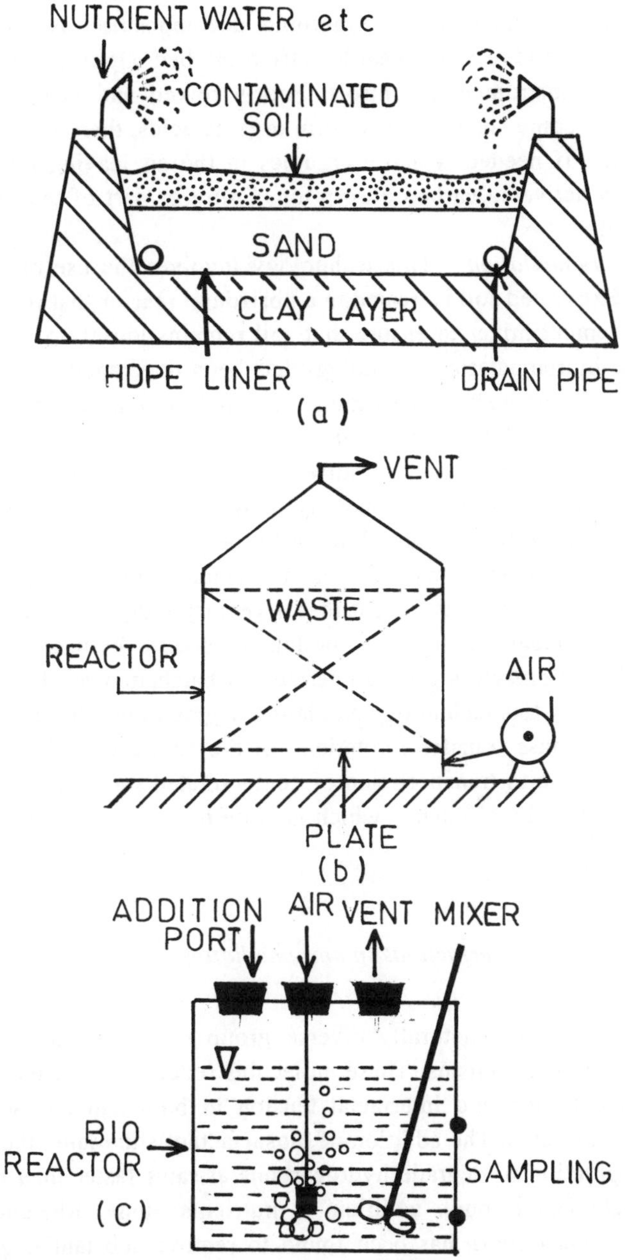

Figure 3 : Ex-situ bioremediation process (a) Land farming (b) composting (c) Slurry bioremediation

reaction rates are higer because aerobic decomposition is an exothermic reaction and there is resricted heat loss from the bed. Optimum temperature is maintained by instrumental controls. Poorly mixed and ventilated compost piles tend to become anaerobic and form organic acids, therefore lime is added to control pH. If needed, volatile organics in the air leaving the bed can be controlled by activated carbon adsorption. Entire or part of the operation can be automated.

c) Slurry bioremediation This technology involves the use of fermenter or bioreactor. Excavated soil is treated in a horizontal reactor that rotates about its axis like a cement mixer to ensure that soil remains loosely packed. Recently slurry phase bioremediation technology has been developed by US-EPA with International Technology Corporation, in which soil in slurry form can be treated either in aerobic, anaerobic or combination of both modes. Slurry is fed either continuously or in batch to stirred tank reactor (Figure 3c). The size of the reactor is dependent on the quantity of waste to be treated as well as the hydraulic retention time required to meet target levels. Suitable microbial environmental parameters with respect to temperature, oxygen and nutrient supply are maintained. In the event that necessary indigenous bacteria are not sufficient to biodegrade organics, pollutant specific cultured microbes are bioaugmented. Bioreactors can be operated in batch, fed batch or continuous culture modes. By this technology maximum degradation rates can be achieved by eliminating mass transfer limitations in the shortest period of time. The additional cost due to transportation of contaminated soil to the treatment site is compensated by its efficiency which is more than double compared to other technologies.

3.4 Application of biosurfactants in bioremediation

Biosurfactants are a structurally diverse group of surface active molecules produced by microorganisms. These molecules reduce surface tension, critical micelle concentration and interfacial tension in both aqueous solutions and hydrocarbon mixtures. The addition of biosurfactant stimulated the indigenous bacterial population to degrade hydrocarbons at rates faster than those which could be achieved through addition of nutrients alone. Rhamnolipid from *Pseudomonas aeruginosa* has been found to remove substantial quantities of oil from contaminated sandy-loam and silt-loam soils. Addition of *P. aeruginosa* UG2 and *Pseudomonas* ML2 biosurfactants to the soil contaminated with hydrocarbon mixture showed stimulatory effect on their degradation by indigenous microbes. Biosurfactants from *Rhodococcus* ST-5, *Bacillus* AB-2 and *P. aeruginosa* SB-30 successfully solubilized and removed hydrocarbons from contaminated surfaces damaged during the oil spill. Biosurfactants also enhance the degradation of many pesticides, poly aliphatic and aromatic hydrocarbons. Biosurfactants have also been found to form complex with heavy metals and desorb them from the contaminated soil. Biosurfactants are biodegradable, non-toxic and can be produced in the natural ecosystems.

4 Biofiltration

Biofiltration (Figure 4) is the technology to remove and oxidize organic gases from contaminated air by passing through a porous column packed with either synthetic or natural sorbents. Table-2 lists the harmful gases and their biodegradability. Contaminated air flows upward through the column, volatile organic compounds (VOC) are adsorbed and then oxidized by microorganisms to carbon dioxide. Biofilter beds also adsorb and oxidize volatile inorganic compounds. The moisture content is maintained at optimum level to support microbial growth without clogging the filter pores. The use of soil and compost as adsorbents in biofiltration technology is gaining significant commercial attention due to following advantages. First, soils have porosities of 40-50% and surface areas ranging from 1 to 100 m^2/g and contain 1-5% organic humic material as coating on the inorganic surfaces. Compost has a 50-80% porosity, similar surface area and contains 50-80% partially humidified organic matter. These porositiees and surface areas are similar to those of activated carbon and other synthetic sorbents. The second advantage of soil and compost is the presence of a microbial population of more than 1 billion microbes per gram, which oxidize organic compounds to carbon dioxide. In addition, moisture in the waste gas stream is beneficial for the microbial oxidation on which the removal efficiency of biofilters depends; moisture in synthetic sorbents, on the other hand, reduces their VOC sorption capacity and removal efficiency. A third advantage is the cost for natural sorbents which is more than hundred times cheaper than the synthetic sorbents.

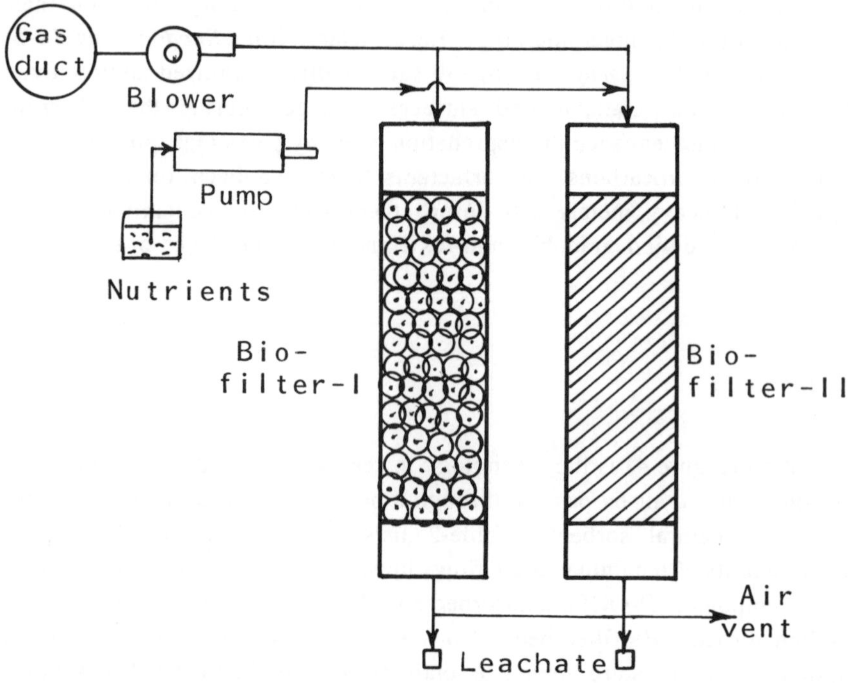

Figure 4 : Biofilteration in removal of toxic substances from gases.

Table 2 : Classification of Gases according to their degradability

I.	Rapidly Degradable Alcohols, Aldehydes, Ketones, Ethers, Esters, Organic acids Amines, Thiols
II.	Rapidly Reactive H_2S, NO_x, SO_2, HCl, NH_3, PH_3, SiH_4, HF
III.	Slowly Degradable Hydrocarbons*, Phenols, Methylene chloride
IV.	Very Slowly Degradable Halogenated hydrocarbons (Trichioroethylene, trichlorethane, pentachlorophenol etc.) Polyaromatic hydrocarbons.

* Aliphatics degrade faster than aromatics such as benzene, toluene, xylene and styrene.

Thus, biofilter is in effect, a mixture of activated carbon, alumina, silica, and lime combined with a microbial population that enzymatically catalyzes the oxidation of the sorbed gases and regenerate the capacity. The oxidation requires no fuel or chemicals and the process can be operated in continuous mode. Gases are inherently more biodegradable than solids and liquids because they are molecularly dispersed, therefore, this technology has gained significant commercialization and over 500 biofilters are operating in Europe and many have been installed in Japan for the removal of odor emission and VOCs like chlorinated hydrocarbons, aldehydes, ketones, aromatics etc. Table 3 summarizes a few commercially operating CPI facilities.

Table 3 : Biofiltration applications in chemical process industries.

US- Company and Location	Application
S.C. Johnson Inc., Racine, WI	90 % removal of Propane and butane, cap. 3,000 cfm.
Monsanto Chemical Co., Springifield, MA	99 % removal of Ethanol and butyraldehyde, cap. 28,000 cfm.
Dow Chemical Co., Midland, MI	Chemical Process gases
Sandoz Basel, Switzerland	Chemical Process gases
Esso of Canada Sarnia, Ontario	Hydrocarbons from storage tanks
Mobil Chem, Co., Canadaigua, NY	Pentane from polystyrene molding
Upjohn Co., Kalamazoo, MI	Odors from a pharmaceutical; cap. 60,000 cfm.

5 Future outlook

With the recent breakthroughs in genetic engineering it is now possible to produce tailor made microorganisms with desired degradative capabilities. Recombinanat DNA technology by-passes all the constraints on genetic exchange between two divergent species. The first report on the application of genetic engineering technology had to do with 'oil-eating' bacteria developed by Anand Chakrbarty.

Eversince the publication of evidence for the transmissibility of drug resistance among the enterobacteria, the importance of extrachromosomal genetic elements or plasmids in the transfer of genetic information from one organism to another has become apparent. The genes responsible for degradation of many compounds have been found to be located on plasmids. Recombination between phenotypically defined plasmids can lead to their co-existence as an aggregate or co-integrate in a single organism. The presence of catabolic plasmids in a bacterial community enables the organisms to share a pool of degradative genes. Inter and intra plasmid recombination may lead to new combinations of genes, thereby, extending the catabolic pathways encoded on the plasmid. The wide variety of catabolic plasmids found in genus *Pseudomonas* provides an explanation for their extensive catabolic capabilities.

Large number of recent evidence suggests that horizontal gene transfer by genetic transformation occurs in the environment. Continuous production and release of DNA in the environment and its long persistence when attached to a solid surface despite presence of DNA ases has been demonstrated. Many species with the potential to make up DNA and propagate its genetic transformation among all major taxonomic groups have been identified. Bacteria are the only organisms capable of natural genetic transformation. Moreover, natural transformation does not require living donor cell as lysed cells produce free DNA. There are about 40-50 species isolated from the natural habitats showing high transformation frequency.

Genetic ecology is concerned with the structural and regulatory genes and how they can be managed *in-situ* to maximize expression. The success of approach as illustrated in Figure 5 is founded in the inherent stability of existing microbial population in the environment, conservation of important genetic information in these organisms and utilizing as and when required. Thus, genetic ecology study identifies physico-chemical parameters that affect both gene amplification and expression in the given environment. It reveals the potential of utilizing genes in microbes and direct chemical processes to decrease pollutant level. It is also used in bioremediation application by amplification of structural genes to degrade pollutants in question through transposition of genes in natural environment. Many operons controlling degradation of pollutants have been found to be associated with transposons.

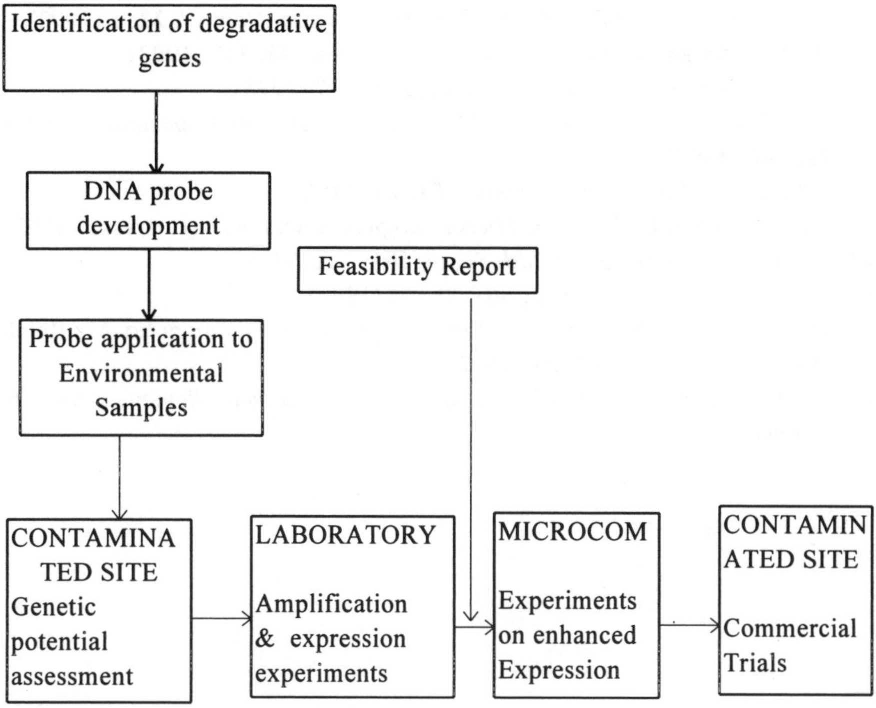

Figure 5 : Application of genetic ecology in commercialization of bioremediation technology.

References

1. H. S. Peavy, D. R. Rowe and G. Tchobanoglous, *Environmental Engeneering.* McGraw Hill. New York (1986).
2. J. D. Desai, and I. S. Bhardwaj, in *Biotechnology, Agriculture and Environment* ed. S. K.Ray, Biotech. Consortium. India Publication, New Delhi, (1995).
3. G. Hamer, T. Egli and K. Mechsner, *J. Appl. Bacteriol.* Sympo suppl., **127**s (1985).
4. J. M. Neff, *Chem. Engg. Prog.*, **83,** 27 (1987).

5. J. Struijs, and R. V. Berg, *Water Res.* **29**, 255(1995).
6. J. F. Tursman, and D. J. Cork, *Crit. Rev. Environ. Control.* **22**, 1 (1992).
7. A. R. Autry and G. M. Ellis, *Environ. Progress.* **11,** 318 (1992).
8. W. F. Ritter, *J. Environ. Sci. Health.,***30A,** 337 (1995).
9. J. A. Rogers, D. J. Tedaldi and M. C. Kavanaugh, *Environmental progress.* **12,** 146 (1993).
10. I. M. Banat, *Bioresource Technol.* **51**, 1 (1995).
11. J. D. Desai and I. M. Banat, *Microbiological Revies.* **61** : (In press). (1997).
12. R. M. Miller, *Environ. Health Persp.* **103**; 59 (1995).
13. H. Bohn, *Chem. Engg. Progress.* **88**; 34 (1992).
14. D. H. Oslon and Y. Tsai in *Environmental Microbiology* ed M. Ralph. Wiley-Liss Inc., New York (1992).
15. M. G. Lorenz and W.Wackernagel. *Microbiological Reviews.,* **58**, 563 (1994).

BIODEGRADATION OF DYES AND DYE EFFLUENTS

K C PATEL AND BHAVIN RAWAL

Department of Biosciences, Sardar Patel University
Vallabh Vidyanagar - 388120, Gujarat, India

1 Introduction

The post world war II growth of the chemical industry has produced millions of tons of hazardous and toxic waste. Large quantities of synthetic materials including pesticides, dyes, dielectric fluids, flame retardants, refrigerants, heat transfer fluids, lubricants, protective coatings and many other chemicals and petroleum products have been produced and they, their residues, and off-grade mixtures have entered the ecosystem. They accumulate in landfills, holding ponds, lakes, rivers, ground waters, soil environments and the atmosphere.

Man is always fascinated by colors. The importance of dyes to human civilization, both ancient and contemporary is well documented. Recent history has shown a switch from natural to synthetic dyes. New dyes and pigments are being introduced into market regularly. The Indian dyestuff industry landscape is changing rapidly bringing new processes and strategies in focus. The rapid change is mainly propelled by shift of the textile and leather industry, which accounts for almost 80% of dyestuff consumption. Both being labour intensive, they have shown a dramatic shift from the developed to the developing countries of Asia-Pacific region. This region today accounts for 42% of the world dyestuff consumption and it is estimated to reach 60% by year 2000[1].

The Indian dyestuff industry has seized this opportunity and transformed itself from an importer to a next exporter of dyes and intermediates. Against a background of low growth in Western Europe and Japan, India is emerging as a new producer of dyestuff and intermediates being third largest exporter among the developing countries after China and Hong Kong. The growth rate of Indian dyestuff exports is at an impressive 25% per year during 1990-1995. India exports 50% of its total production estimated at 100,000 tons in 1995 to major market being Europe and USA.

The Indian dyestuff industry has remain highly fragmented consisting of organized companies and small scale units. This fragmented structure has done lot, for uplift of nations economy by providing low cost, versatile products. On other hand it has constrains like product standardization and consistency of products.

Productivity, conservation, safety, pollution control and health are the basics of the sustainable industrial growth. The Industrial development will suffer a serious set back, if any one of these factor is neglected[2]. At least in developing countries, the pollution control was given the least priority till recently. Initially, pollution control measures were considered to be an undue burden on the industry. Subsequently, it was realized that pollution control is an essential measure for healthy environment without much direct benefit to industries concerned.

In dye and dyestuff manufacturing industries most neglected part at industrial sites are the effluent treatment plants. Most of the plants are working inefficiently and are releasing very high amount of dyes and dyestuff in natural ecosystem. Estimates indicate that 12% of the synthetic textile dyes used yearly are lost to waste stream during dyestuff manufacturing and textile processing operations. Approximately 20% of these losses enter the environment through effluents from the waste water treatment plant[3].

In Gujarat, a state located in Western India, Dye manufacturing units are in different industrial estates of the state. Vatwa near Ahmedabad, Nandesari near Baroda and Ankleshwar near Surat are the major industrial areas where small scale industries are located. For them also the major problem is the effluent treatment.

A survey carried by Sudarshan Chemical Industries, Pune for Green Environment Service Cooperative Society Limited (GESC), a society established by Gujarat Dyestuff Manufacturing Association, Ahmedabad, revealed that major problem causing pollutants are vinyl sulphone, H-acid, direct dyes, reactive dyes, acid dyes, turquoise blue and other intermediates. Quantity of effluent generated by various units vary from 2-20 m^3/day. Most of the units have batch operations and resulted effluent was highly acidic with pH ranging from 0.01-9.00. Most of the effluent was found highly colored (Pt-Co. unit not available). They were having very high chemical oxygen demand (88,000 ppm) and biological oxygen demand (15,250 ppm). TOC, TDS and TSS were also found to be beyond the permissible range.

2 Toxicity of dyes and Intermediates

With few exception, the normal presence of organic colorants in the environment possess the problem in terms of acute ecological effects[4]. Certain dyestuff exhibit toxic effect towards microbial population and can be toxic and/or carcinogenic to animals[5,6]. The possible contamination of drinking water supplies is also of concern as certain class of dyes are known to be enzymatically degraded in the human digestive system, producing carcinogenic substances[7,8,9,10].

Acute, or short term toxic effects are generally well known and are controlled by keeping concentration of chemicals in the work place atmosphere below prescribed limits and avoiding physical contact with the material. Chronic effect on the other hand frequently does not become apparent until after many years of exposure. Higher incidences of bening and malignant tumors, especially in the bladder of workers exposed to certain intermediates and dyes were recorded in dye producing countries during period 1930-1960. The specific compounds involved were 2-napthyl amine, 4-amino biphenyl benzidine, fuschsine and auramine. There are considerable evidences which shows that metabolites of these compound are the actual carcinogenic agents[11]. Other actual or suspected carcinogenic agents such as the nitrosoamines, or N-nitroso compounds[12], polycyclic hydrocarbons, alkylating agents and other individual compounds such as the dichromates, should be considered in the wider context of environmental concern rather than as dye intermediates. The positive link between benzidine derivatives and 2-napthyl amine with bladder cancer prompted the introduction of stringent government regulation to minimize such occurrences in the future. There are reports indicating that health hazards commonly associated with the dyestuff industry are urinary bladder cancer due to aromatic amines, cyanosis leading to coma, death due to methaemoglobinemia, skin diseases, respiratory ailments and acid/alkali burns[13].

Uncontrolled discharge of raw or semitreated effluent into inland water bodies may cause serious degradation of water quality leading to fish mortality. Effect of industrial effluents on growth and behaviour studies of fish are reported[14,15]. Growth of the fish *Cyprinus carpio* is affected because of decrease in food consumption in presence of dyestuff effluent[16]. Oxidative enzymes like succinate dehydrogenase, lactate dehydrogenase and tissue respiration is also affected by dye containing textile industry waste in this fish[17].

When plants like *Wolffia arrhiza*, and *Spirodella polyrrhiza* a small free floating plant, were grown in effluent containing phenol and methyl violet, there was uptake of violet colour by leaves and wilting occurred. After 72 hours of treatment no plant survived, where as in control there was survival of both species[18].

Apart from the toxicity of dyes or the intermediates present in the effluent and their adverse environment effects another aspect which emphasis the need for treating such effluent is a probable "water crunch"[19]. Both require the effective treatment of dye effluents for environmental safety and reuse of the water for other processes.

3 Conventional Effluent Treatment Methods

Methods of Effluent treatment for dyes may be classified broadly into three main categories: Physical, Chemical and Biological (Table 1).

Table 1: Methods for treatment of dye containing effluents.

Physical	Chemical	Biological
Adsorption Sedimentation Floatation Flocculation Coagulation Foam Fractionation Polymer Flocculation Reverse Osmosis Ionization radiation Incineration	Neutralization Reduction Oxidation Electrolysis Ion exchange Wet air oxidation	Stabilization pond Aerated lagoons Trickling filters Activated sludge Anaerobic digestion Bioaugmentation

The conventional effluent treatment methods can be divided into four stages:

i) Preliminary ii) Primary iii) Secondary iv) Tertiary

Preliminary treatment processes of dye effluents include homoginization, neutralization of pH and disinfection. Primary stages includes physical phenomenon such as screening, sedimentation, floatation and flocculation to remove debris, undissolved chemicals and particulate matter. In secondary stage organic load is reduced by combination of physico-chemical separation and biological oxidation. Tertiary stages are important because they serve as a polishing of effluent. These methods includes adsorption, ion-exchange, chemical oxidation, hyperfilteration (reverse osmosis) and other processes.

4 Biological methods

Considering the problems associated with the physical and chemical processes, biological methods gained ground in treating the highly polluted industrial effluents to improve their quality for reuse or recyling. These methods are economical, require less or no energy and may lead to valuable end products. A few examples of such systems are earthworm, fish farming, biomethanation and microbial degradation. Higher forms of biosystems like earthworm and fishes do not tolerate high COD and BOD. They can tolerate hardly 500 to 1000 ppm BOD and require large quantities of dilution water so are suitable for treatment of low strength industrial effluents. Lower form of life like micro-organisms have better tolerance and can grow at higher BOD values[19].

4.1 Use of Microorganisms

Microorganisms are a wonderful creation of nature. They can degrade almost any thing on this planet. It is a question of time taken by them to degrade various materials[21]. Micro-organisms can withstand BOD as high as 40,000 to 50,000 ppm and they can degrade a variety of organic and inorganic pollutants[20].

They play a major role in the biodegradation of dyes and intermediates during the biological treatment of effluents.

Biological methods using biodegradation capabilities of microorganisms can be classified into two groups: Aerobic and Anaerobic methods. The four common aerobic biological treatment processes are :

1) Activated sludge 2) Aerated lagoons 3) Stabilization ponds
4) Trickling filters

Even if we achieve the desired BOD and COD by various biological treatments, color will still remain. This is not aesthetically acceptable in case of distillery, as well as paper and pulp effluent, and color may be natural as in the case of distillery spent wash (caramel and lignin) or synthetic as in textile effluents. Their removal by chemical treatment is highly expensive. Although efforts of biological color removal are on, there is no head way made so far. There are some reports of color removal from paper and pulp mill effluent using fungi like *Coriolus versicolor, Polyporus versicolor, Phanerochaete chrysosponum*[21,22].

There are reports of effective treatment of dye industry waste in a laboratory scale activated sludge process[23]. Effluent containing aniline, phenol, methyl violet and rhodamine B, was having 5576 $mg.l^{-1}$ of COD, 896 $mg.l^{-1}$ of total organic content and 31.5 $mg.l^{-1}$ phenol content. *Pseudomonas alcaligenes* and *P. mendocina* present in the sludge showed reduction of 60% in COD, 37% in TOC and 92% in phenol content. Decrease in the optical density of the color from an initial 0.915 to .360 at 580 nm was also observed.

Thus, the biological treatment, more precisely microbiological treatment is the main hope for future treatment technologies.

5 Microbiology of Dye Degradation

Dyes are biodegradable by only few micro-organisms. Table-2 gives an overview of the group of microbes reported to degrade dye by metabolizing or significantly altering their structure.

Table 2: Dye degrading organisms

Microbial classification and examples (genera)	Enzyme/ pathway	Dyes used for degradation	Methods used to determine dye degradation
Actinomycetes			
Streptomyces	Peroxidase	Azo dyes	Spectrophotometric
Red Yeast			
Rhodotorula	Oxidative mechanism	Crystal Violet	Spectrophotometric
Bacteria			
Pseudomonas	Cell free	Azo,	Spectrophotometric

	extract	Triphenyl-methane Anthraquinone	
Aeromonas	Cell free extract	Azo dyes	G.C., M.S. Spectrophotometer Flurometry
Flavobacterium	Lignin Peroxidase	Azo dye	Spectrophotometer
Fungi			
Basidiomycetes Coriolus Phanerochaete Pleurotus	Lignin Peroxidase MnP	Azo dye Copper Pthalocy-nine Crystal Violet Orange II Azo and heterocyclic Triphenyl-methene	Spectrophotometer
Geotricum	----	Various dyes	Spectrophotometer

5.1 Bacteria

Biodegradation of azo dyes by anaerobic and aerobic micro-organisms has been studied extensively and involvement of both the reductive and oxidative pathways have been proposed[24,25]. Azo reductase has been detected in some bacteria[26]. This enzyme reductively cleaves the azo linkage of the dyes. Report are available showing degradation of azo dyes using cell free extract of *Aeromonas hydrophila* var 24B, an azo dye assimilating bacteria[27]. Degradation of compounds like 4-aminobenzene, 4-dimethylamino benzene, with aniline as a product 2-carboxy-4-dimethyl amino benzene and 2-methoxy carbonyl-4-carboxyazobenzene were not degraded. They have proposed the involvement of reductase in the dye degradation. An overall oxidative pathway for azo dye degradation by *Pseudomonas cepacia* has also been proposed in which first reaction is catalyzed by azo reductase. Thus ability of most bacteria to decolorize azo dyes involves the initial reductive cleavage of azo linkage. While reduced products; aromatic amines are essentially non degradable by anaerobic micro-organisms, may further oxidize the reduced product via deamination and/or hydroxylation reaction[28]. There are reports[29] of oxidative cleavage of azo dye by *Flavobacterium* peroxidase. This bacterium was isolated as a pentachlorophenol degrading organism from the soil. Comparision of azo dye 3,5-dimethyl-4-hydroxyazo-benzene-4' sulfonic acid oxidation with native and heat inactivated enzyme preparations and with and without H_2O_2 confirmed the involvement of an extracellular peroxidative enzyme in azo dye decolorization. The presence of syringyl or guaiacyl groups in the dye structure results in enhanced oxidation by peroxidase of bacteria and streptomyces. These shows that azo dyes can be made more reactively oxidizable with peroxidase by appropriately modifying the substitution patterns of their aromatic ring.

The position of hydroxy groups are also important in biodegradation of azo and triphenylmethane dye[30]. The hydroxy group in para position with respect to azo group, increased the decolorization of the dye. Naphthalene ring containing dye was rapidly decolorized compared to benzene ring containing dye. Decolorization of triphenylmethane dye was found inferior to that of azo dyes. Triphenylmethane has been reported to be inhibitory to the growth of *Bacillus subtilis*[31]. Considering the capacity of the bacteria for decolorization of dyes, specific dye assimilating bacteria can be added to the activated sludge for the treatment of dye containing effluents.

5.2 Actinomycetes

They form a large group of indigenous soil bacteria. Degradation of cellulose and lignin by Streptomyces spp. is well documented[32,33,34]. Lignin peroxidases are apparently involved in lignin degradation by this branching bacteria[35]. Extracellular peroxidase appear to be involved in the degradative mechanism. Studies suggest that azo dye degradation by *Streptomyces* spp. depends on specific chemical structure of dyes and involves peroxidase catalyzed reaction[36,37,38]. Addition of guaiacol, via an azo linkage, onto the commercial acid yellow 9' led to its significant degradation by some streptomyces strain whereas none could degrade the original dye[39]. Thus modified azo dyes containing lignin like structures were found to be degraded more efficiently by

Streptomyces[40]. Oxidation of azo dyes by *Streptomyces chromofuscus* A11 involves an extracellular peroxidase similar to manganese dependent peroxidase and horse raddish peroxidase. Substitution on the ring structure is important when dyes are to be degraded enzymatically. The structural pattern of the hydroxy group in para position relative to the azo linkage and of two methyl substitution groups in ortho position relative to the hydroxy group was the most susceptible to degradable. Replacement of the sulfonic group with a carboxylic group enhanced overall dye degradation by *S. chromofuscus* A11[41].

5.3 Fungi

The white rot group of fungi comprises of wood degrading Basidiomycetes and few Ascomycetes, which have in common, capacity to degrade lignin as well as other wood components. They possess the ability to produce extracellular enzymes which oxidize phenolic components including many related to lignin.

Research efforts are being done world wide to apply white rot fungus on environmental pollutant both in soil and in liquid effluents. Most of the research studies are on *Phanerochaete chysosporium* which is widely used Basidiomycete as it is is exceptionally versatile at degrading xenobiotics. It produces a variety of non-specific ligninases, which also act upon organic molecules other than those present in lignin[42].

Following are the advantages of the use of white rot fungi in biodegradation of hazardous wastes[43].

1. White rot can be grown on cheap substrates such as forest or agrowaste.
2. Given effective engineering system these fungi can be used to treat liquid effluents and soil.
3. White rot enzymes themselves may be used directly, particularly after immobilization on a suitable carrier.

6 The RGE system

A promising new area in environmental biotechnology is the use of Radical Generating Enzyme (RGE) system to degrade pollutants. Decolorization of polymeric dyes by *P. chryosporium* has been reported[44]. The degradation was suppressed by nitrogen sufficiency and required oxygen and was blocked by inhibitors such as azide, thiourea and potassium cyanide. Decolorization of Poly Blue (poly vinyl amine sulfonate anthroquinone) is reported with the fungi known to degrade lignin[45]. These findings confirms the role of lignolytic enzymes in dye decolorization. The role of RGE system was established by applying ligninase together with H_2O_2 and manganese ions to a waste water containing o-cresol, 99% of which was removed due to radicals which were generated. This study suggested the possibility of developing a practical system based on the use of this fungal enzyme to detoxify chemicals in wastes[46]. Decolorization of culture medium of *P. chrysosporium* containing azo and heterocyclic dyes particularly Orange II, Tropaeolin O and Azure B and to lesser extent Congo red was observed, at concentration of 57, 63, 16 and 76 μ mole respectively. After 24 hours incubation over 90% of the initial color from all four dyes disappeared under nitrogen limited condition and reached upto 95% after 48 hours. Adsorption was an important factor under nitrogen sufficient condition. Spectrophotometric examination of methanol extract of fungal mats after 5 days incubation indicated binding of 11% Orange II, 18% Tropaeolin O, 14% Azure B and 49% Congo red to the mycelia. No dye was detected in extract from nitrogen limited cultures except incase of congo red (6% bound). This findings indicated that biodegradation of dye occurs extensively under nitrogen limited condition. Both biodegradation and adsorption to the fungal mycelia are important processes in removing dyes from the incubation media of *P. chrysosporium*. So commercial application of this treatment will enable to take advantage of both these dye removal processes[47].

The triphenylmethane dye crystal violet (N,N,N',N',N",N"-hexamethyl para rosaniline) is routinely used in medicine and as a stain in textile dye. Its persistance in nature has also been reported. This class of chemical is responsible for tumor growth promotion in some species of fish. Its slow biodegradation has been attributed to its toxicity to a range of micro-organisms. Using initial crystal violet concentration of 12.3 mM it was found that *P. chrysosporium*, N-demethylated the dye at each of three nitrogenous in the dye molecules and in turn total decolorization. Degradation products were identified as N,N,N',N',N"-penta, N,N,N',N"-tetra and N,N',N"-tri-methyl pararosaniline. Biodegradation appears to proceed beyond N,N',N"-tri-methyl pararosaniline, as evidenced by the fact that two additional metabolites were found by HPLC. Extracellular fluid from the culture and purified peroxidase also carried out N.demethylation in presence of H_2O_2[48].

Dealkylation of N,N-dimethylaniline has been shown by horse raddish peroxidase[50] and chloroperoxidase[49] and it was observed that the lignin degrading system is acting non-specifically in this particular case, for the initial attack on the nitrogen atom, which is absent in lignin itself. Crystal violet bears little similarity to substructure proposed to comprise lignin.

The decolorization of six other triphenylmethane dyes by lignolytic cultures and purified ligninase emphasise their role in biodegradation of dyes[48]. The initial oxidation of brilliant green, malachite green and ethyl violet, which contains N-alkyl group may proceed in a similar fashion to that of crystal violet. Those without such groups, like parasomaline, cresol red and bromophenol blue must be degraded by different mechanisms.

Study[51] with radio labelled ^{14}C found that under nitrogen limited condition *P. chrysosporium* extensively mineralized ^{14}C ring labelled azo dyes 4-phenyl azo phenol, 4-phenyl-azo-2-methoxyphenol, disperse yellow 3,[2-(4'-acetamidophenyl azo)-4-methyl phenol], 4-phenyl azo-aniline N,N-dimethyl 4-phenylazoaniline, disperse orange 3[4-(4'-nitrophenyl azo)-aniline] and solvent yellow 14(1-phenyl-azo-2-napthol). Twelve days after addition to cultures the dyes were mineralised 23.1% to 48.1%. Aromatic rings with substituents such as hydroxyl, amino, acetamido, or nitro functions were mineralized to greater extent than unsubstitutes rings. Most of these dyes were degraded extensively only under lignolytic conditions. However, 4-phenyl azo phenol and 4-phenyl azo methoxy phenol were mineralized to a lesser extent under nitrogen sufficient condition as well.

Addition of guaiacyl substructure onto azo dyes, enhances dye decolorization by *P. chrysosporium* and *Streptomyces*[39]. *P. chrysosporium* was found superior in decolorization. Addition of guaiacol into dye structure makes it conjugated multiunsaturated system. This makes possible to change only one fragment of the molecule and yet have the entire conjugated system becoming accessible to enzymatic attack, particularly when microorganisms use oxidative enzymes that generates cation radicals.

Isoenzymes of lignin peroxidase H_2 and H_7 were shown to require veratryl alcohol to catalyze the reaction whereas isoenzyme H_8 considerably mineralized the dyes in absence of veratryl alcohol[52].

There are reports[53] of some isolates which have capacity to decolorize wide range of structurally dissimilar synthetic dyes. Most interesting finding was that the dye decolorization by *P. chrysosporium* was minimum as compared to other isolates. This finding may encourage the scientists to search potential degraders which may be more efficient than *P. chrysosporium*. The results obtained by them were preliminary and media used was unoptimized. Better degradation can be obtained under optimal condition not only in rate and extent of decolorization, but also with respect to the range of dyes to be degraded.

Geotrichum candidum Dec. 1. was found to decolorize 18 kind of reactive, acidic and disperse dyes and 3 model compounds[54]. Organisms is a yeast like fungus which could be very important in bioremediation processes due to two reasons:

1. It has capacity to decolorize various array of dyes in soil and liquid medium.
2. Organisms were fast growers as compared to other white rot fungus and decolorization was obtained in hours. While with other organisms decolorization was obtained in days. However detailed understanding of dye degradation by Dec 1 is required in order to establish fungus base treatment systems for clearing up of dye industry effluents and for bioremediation of dye contaminated soil.

Laccase oxidation of phenolic azo dyes has also been studied[55]. Only the substituents like 2-methyl, 2-methoxy, 2,3-dimethyl, 2,6-dimethyl, 2,3-dimethoxy and 2,6-dimethoxy on phenol served as substrate. This study revealed that laccase oxidation can result in the detoxification of azo dyes.

7 Cell and enzyme immobilization for dye decolorization.

There are reports of treatment of phenolic effluents with immobilized white rot fungi, that removes colour and degrades toxic compounds[56,57]. The mycellial color removal (MyCoR) process uses *Phanerochaete crysosporium* mycelium immobilized on rotating biological contactors[56,58] and requires an initial growth period after inoculation prior to exposure to effluent. *Coriolus versicolor* immobilized in calcium alginate beads[59] or as mycellial pellets[57] has been used in air lift reactors, but there are diffusion limitations and alkaline effluents have to be adjusted to pH 4-5 before treatment.

The laccase of *Neurospora crassa* has been immobilized on concanavalin A-sepharose and on cyanogen bromide activated sepharose 4-B[60]. Laccase of *Coriolus hirsutus* has been immobilized on polyacrylonitrile[61]. These supports, however, are not suitable for application in nature due to their instability and susceptibility to microbial attack. Laccase of *Rhizoclonia praticola* has been immobilized on celite[62], but the immobilization did not significantly increase the stability of the enzyme compared to the free enzyme. Porous glass has been claimed to be superior support for enzyme immobilization because of its structural stability and large surface area[63, 64]. It is widely used for covalent coupling as the -OH groups on the surface can be linked to enzyme via salienization[65]. Leonowicz et al.,[66] immobilized laccase of *Trametes versicolor* on porous glass beads that were activated with 3-aminopropyltriethoxysilane and gluteraldehyde. The support immobilized 100% of the enzyme with 90% of the activity retained. Porous glass has not been found suitable for industrial purpose due to high price and lack of stability in continuous flow system[67].

Our laboratory data (unpublished) shows and supports the potential ability of white rot fungus to decolorize the commercial dyes. Study included the use of white rot fungus *Antrodiella* RK-1, a phenol oxidase secreting fungus for the decolorization of pigment, copper phthalocyanine and azo dyes. Correlation with laccase production and copper phthalocyanine degradation in the medium has been observed (Table 3). Effective decolorization upto 83% of the effluent was found when inoculated with the fungus. The treatment also helped in reducing the toxicity of the effluent towards seed germination and growth (Table 4). Seeds were soaked for 18 hrs in distilled water, effluent and effluent obtained after fungal growth for 20 days. Germination was recorded after 48 hrs and shoot length of the germinated seeds was measured after 168 hrs.

Table 3 : Laccase production and dye decolorization in medium containing 1% glucose and 100 ppm copper phthalocyanine

Days	Laccase U/ml	% of colour removal
3	0.19	42
6	0.49	53
9	2.06	71.6
12	7.67	100

Table-4: Effect of treated and untreated effluent on germination and growth of cereals and pulses.

Treatment	Rice		Wheat		Green grams		Bengal grams	
	24 hrs	168 hrs	24 hrs	168 hrs	24 hrs	168 hrs	24 hrs	168 hrs
Distilled water	14	4.1	11	8.6	15	9	7	4
Effluent	1	--	0	--	0	--	0	--
Treatment effluent	15	4.5	15	7.6	15	8.8	4	2.5

24 hrs : no. of seeds germinated out of 15
168 hrs : shoot/root length in cm. of average 5 seeds, - not grown

8 Conclusion

Thus the micro-organisms and their enzyme system have great potential in decolorization of dye and dye containing effluents. Incorporation of the enzymes and the specific organisms in the dye removal strategies along with conventional processes may accelerate color removal. Efficiency and stability of both can be improved further by immobilization on various carriers.

References

1. K. Purohit, in Economics Times, Dyes and Intermediates, special focus, Ahmedabad Edition (7 Aug, 1996).
2. K. P. Nyati, Paper presented in the workshop on Environmental Statement, Bangalore, (1995).
3. A. Reife, in Kirk and Othmer Encyclopeida of Chemical Technology. John Wiley and Sons, New York, (1995).
4. E. A. Clarke and R. Anlinker, *Rev. Prog. Color* **14**, 884 (1984).
5. G. B. Michaels and D. L. Lewis, *Env. Toxicol. Chem.* **4**, 45 (1985).

6. K. T. Chung, G. E. Fulk, and M. Egan, *Appl. Env. Microbiol.* **35**, 558 (1978).
7. C. T. Helmes, *J. Env. Sci. Health* A **19**, 97 (1984).
8. C. C. Sigman, *J. Env. Sci. Health* A **20**, 427 (1985).
9. R. A. Levine, W. L. Oller, C.R. Nony and M.C. Bowman, *J. Anal. Toxicol.* **6**, 157 (1982).
10. C. R. Nony, M. C. Bowman, T. Cairns, L. K. Lowry and W. P. Tolos, *J. Anal. Toxicol.* **4**, 132 (1980).
11. W. C. Hueper, in *Occupational and Environmental Cancers of the Urinary system Yale Univ.* Press, New Haven., (1969).
12. C. E. Surls, in ACS Monograph vol. 173. *American Chemical Society. Washington,* DC. 491 (1977).
13. A. S. Menon, A. K. Agarwal, N. Mohan Rao, S. K. Kashayp and S. K. Chatterjee. Report on the Survey by National Institute of Occupational Health. Ahmedabad, India.
14. W. K. Seim, J. A. Lichatowich, R. H. Ellis and G. E. Davis, *Water Research,* **11**, 189 (1977).
15. P. W. Webb and J. R. Brett, *J. Fish Res. Bd. Can.* **29**, 1955 (1972).
16. M. Sakthivel and K. Sampath, *Ind. J. Expl. Biol.* **27**, 1032 (1989).
17. M. Sakthivel, K. Sampath and T. J. Pandian, *Ind. J. Expl. Biol.* **29**, 979 (1991).
18. Pragnya Kanekar, M. S. Kumbhojkar, Vinaya Ghate and Seema Sarnaik, *Journal of Environ. Biol.* **14**, 129 (1993).
19. T. N. Bhavanishankar in Microbes for Better Livings, MICON 1994, AMI conference, 219, (1994).
20. R. K. Guthrie and E. M. Davis, in, *Advances in Bitoechnological Processes.* Alan R. Liss. Inc., (1985).
21. R. L. Crawford and D. L. Crawford, *Enz. Microb Tech.* **2**, 11 (1980).
22. R. L. Crawford and D. L. Crawford, *Enz. Microbe Tech* **6**, 434 (1984).
23. Pragnya Kanekar and Seema Sarnaik, *Environmental Pollution.* **70**, 27 (1991).
24. K. T. Chuny and S. E. Stevens, *Jr. Crit. Rev. Micorbiol.* **18**, 175 (1992).
25. T. K. Kulla in, *Microbial degradation of Xenobiotics and Recalcitrant compounds.* eds.T. Leisinger, A.M. Cook, R. Hutter and R. Neusch Academic press. London. (1981).
26. T. Zimmermann, F. Gasser, H.G. Kulla and T. Leisinger. *Arch. Microbiol.* **138**, 37 (1984).

27. C. Yatome, T. Ogawa, K. Itoh, A. Sugiyama and E. Idaka, *Journal of Society of Dyers and Colorists.* **103**, 395 (1987).
28. E. Idaka, T. Ogawa, H. Horits, *Bull. Env. Contam. Toxicol.* **39**, 100 (1987).
29. W. Cao, B. Mahadevan, D.L. Crawford and R.L. Crawford, *Enz. Microb. Tech.* **15**, 810 (1993).
30. C. Yatome, T. Ogawa, D. Koga and E. Idaka, *Journal of the society of Dyers and colorists.* **97**, 166 (1981).
31. T. Ogawa, M. Shibata, C. Yatome and E. Idaka, *Bull. Env. Contam. Toxicol,* **40**, 545 (1988).
32. D. L. Crawford, in, *Actinomycetes in Biotechnology* eds.M. Goodfellow, S. T. Williams, M. Mordarski, Academic Press, San Diego, (1988).
33. L. A. Deobald and D. L. Crawford. *Appl. Microbiol. Biotechnol.* **28**, 158 (1987).
34. M. B. Pasti and M.L. Belli, *Appl. Env. Microbiol.* **56**, 2213 (1990).
35. M. Ramchandra, D. L. Crawford and A. L. Pometto, *Appl. Env. Micorbiol.* **53**, 2754 (1987).
36. M. B. Pasti-Grigsby, A. Paszczynski, S. Goszczynski, D. L. Crawford and R. L. Crawford, *Appl. Environ. Micorbiol.* **58**, 3605 (1992).
37. M. B. Pasti, S. Goszczynski, W. Cao and Z. Wang, Abstr. Gen. Meet American Soc. Micorbiol. New Orleans, L.A. 153, (1992).
38. A. Paszczynski, M. B. Pasti, S. Goszczynski, D. L. Crawford and R. L. Crawford, Abstr. Int. Symp. on Appl. of Biotechnology to Tree Culture Protection and Utilization. Columbus, OH (1991).
39. A. Paszczynski, M. B. Pasti, S. Goszczynski, D. L. Crawford and R. L. Crawford, *Enzy. Microb. Tech.* **13**, 378-384 (1991).
40. A. Paszczynski, M. B. Pasti-Grigsby. S. Goszczynski, R. L. Crawford and D. L. Crawford, *Appl. Env. Microbiol.* **58**, 3598 (1992).
41. M. B. Pasti-Grigsby, N. S. Burke, S. Goszczynski, R. L. Crawford, *Appl. Env. Microbiol.* **62**, 1814 (1996).
42. J. Morton and A. M. Stern, *Tappi Jour.* **52**, 1975 (1969).
43. M. Wainswright, in, *An Introduction to Fungal Biotechnology.* John Willey and Sons (1992).
44. J. K. Glenn and M. H. Gold, *Appl. Env. Microbiol.* **45**, 1741 (1983).
45. M. W. Platt, Y. Haler, H. Chet, *Appl. Microbiol. Technol.* **21**, 394 (1985).
46. M. D. Aitkin, R. Venkatadri and R. L. Irvine, *Water Research,* **23**, 443 (1989).
47. C. Cripps, J. A. Bumpus, S. D. Aust, *Appl. Env. Microbiol.* **56**, 1114 (1990).

48. J. A. Bumpus and B. J. Brock, *Appl. Env. Micorbiol.* **54**, 1143 (1988).
49. G. L. Keddris, D. R. Koop and P. F. Hollenberg, *J. Biol. Chem.* **225**, 10174 (1980).
50. G. L. Keddris and P. F. Hollenberg, *J. Biol. Chem.* **258**, 8129 (1983).
51. J. T. Spadaro, M. H. Gold and V. Ranganathan, *Appl. Env. Microbiol.* **58**, 2397 (1992).
52. P. Ollikka, Kirsi Alhonmaki Velimatti Leppamen, Tuomo Glumoff, Timo Raijola and Ilari Suominen, *Appl. Env. Microbiol.* **59**, 4010 (1993).
53. J.S. Knapp, P.S. Newby and L.P. Reece, *Enz. and Microbial Technolgoy,* **17**, 664 (1995).
54. S.J. Kim, K. Isikawa, M. Hirai and M. Shoda, *Journal of Fermentation and Bioengineering,* **79**, 601 (1995).
55. M. Chivukula and V. Ranganathan. *Appl. Env. Microbiol.* **61,** 4374 (1995).
56. D.C. Eaton, H.M. Chang, T.W. Joyce, T.W. Jeffries and T.K. Kirk, *Tappi Jour*, **65**, 89 (1982).
57. G. Royer, M. Desrochers, L. Jurasek, D. Rouleau, and R. C. Mayer, *J. Chem Technol. Biotechnol.* **35B**, 14 (1985).
58. G. Sundman, T. K. Kirk, and H. M. Chang, *Tappi Jour,* **64**, 145 (1981).
59. G. Royer, D. Livernoche, M. Desrochers, L. Jurasek, D. Rouleau and R.C. mayer, *Biotechnol Lett.* **5**, 121 (1982).
60. S. C. Froechner, and K. E. Eriksson, *Acta Chem Scand* B **29**, 641 (1975).
61. N. A. Goncharova, V. P. Gavrilova, I. I., Shamolina, A. B. Lobora, L. A. Vol, *Appl. Biotechnol. Microbiol.* **21**, 262 (1985).
62. K. L. Shuttleworth and J. M. Bollay. *Enzyme Microb Technol.* **8**, 171 (1986).
63. R. A. Messing. *Enzymologia.* **39**, 12 (1970).
64. H. H. Weetall, *Science,* **166**, 615 (1969).
65. H. H. Weetall. *Methods Enzymol.* **44**, 134 (1976).
66. A. Leonowicz, J. M. Sarkar, J. M. Bollay. *Appl. Microbiol. Biotechnol.* **29**, 129 (1988).
67. J. F. Kennedy, C. A. White, and A. Wiseman, in *Handbook of Enzyme Technology,* Ellis Hardwood. Chichester, U.K. (1985).

ALGAE OF DAMAGED ECOSYSTEMS - SOME CASE STUDIES FROM SOUTH INDIA

V. SANKARAN

*Department Of Botany, N.G.M College Pollachi - 642 001,
Tamilnadu - India*

1 Introduction

Industrialization and urbanization are a necessary evil at the regional, national and global levels. They damage the ecosystem to varying degrees. In South India also, both the edaphic and aquatic systems are polluted by industrial and urban pollutants. Even in such damaged ecosystems, exist various algal communities, tolerating and indicating pollution. These algal communities not only indicate pollution, but also indicate its reversal. Algal compositions of certain specific, damaged ecosystems-both edaphic and aquatic (fresh and brackish waters) are brought out here. The materials and methods and the discussions concerned are given under each case study.

2 Case Study I : Cement Dust Polluted Soil

Sankaran[1] has investigated the algal flora of soils situated near a cement factory near Pollachi, Tamil Nadu. The soil receives ample cement dust and is very adverse for survival of biotic communities.

The A.C. Cement factory at Madukkarai adjoining Coimbatore is 40 km Northwest of Pollachi. The cement dust from the kiln of the factory is dissipated across a radius of a few kilometres in all directions throughout the year, depending upon the direction of wind. As a consequence, cement dust is deposited on the soil around and also far away from the factory.

2.1 Material and Methods

Soil samples from three different localities, South of the factory were collected at random, in polythene bags at three different levels viz., (i) Surface, (ii) 15 cm below surface, and (iii) 30 cm below surface, with the help of a clean spatula, brought to the laboratory and placed in sterile glass jars. Sterile, deionized water was added to these soil samples and incubated in light for 18 hrs. (fluorescent lamps 2 x 40 W + 20^0 C) /day for a period of two months. Samples were kept in triplicate. In about 10 days, the water above the soil in the samples turned greenish and contained algal growth. This was periodically withdrawn under sterile conditions from the jars and the algal forms were identified[2].

2.2 Results and Discussion

The algal forms that have been observed from the polluted soil are listed in Table 1.

Table 1 : Algal forms in the Polluted Soil

Name of Algal	Surface	15 cm below surface	30 cm below surface
1. *Lyngbya allorgei*	+	-	-
2. *Phormidium Sp*	-	+	-
3. *P. luridum*	+	+	+
4. *P. cebennense*	-	-	+
5. *Microcoleus acutissimus*	+	-	-
6. *Nostoc commune*	+	-	-
7. *N. punctiforme*	-	-	+

\+ indicates presence of alga
\- indicates absence of alga

The soil polluted with cement offers a very adverse environment for plant growth because of its physical properties[3]. But, this polluted soil is fairly rich in algal flora, and contains only filamentous blue-green algae. The coccoid blue-green algae and members of other algal groups are absent from this soil. The ability of algae, particularly blue greens to survive under adverse conditions is well-known. They have been shown to tolerate extremes of temperature, desiccation and salinity[4]. The blue-green algal genera *Lyngbya, Phormidium, Microcoleus and Nostoc,* whose species have been recorded during the present investigation have been reported by Cameron[5] to occur in desert soils of South Western Unites States. The Species of *Nostoc* occurring here are known for their Nitrogen fixing ability[6]. Though it is not known whether the forms recorded here are actually Nitrogen fixers, their potentiality to fix N_2 has to be considered. Experimental investigative approach will be worth while to find out ways and means of increasing the fertility of the polluted soils by promoting the growth of these potential N_2 fixers mentioned, so that attempts could be made to reclaim the land from hazards of pollution.

3 Case Study II : Soil Ecosystem From an Industrial Area Near Pollachi

Sankaran etal[7] have analysed the algal flora of a soil receiving coconut husk waste and electroplating wastes. This edaphic ecosystem is an example of another highly damaged ecosystem. The Industrial Estate situated at the eastern outskirts of Pollachi, discharge coconut husk waste and electroplating waste on to the land from a defibering unit and electroplating unit respectively. The Electroplating works, discharge the waste onto the land area. The coir factories (defibering units) dump the coconut husk into water and release the water into drainage. The algal flora from these two spots receiving the industrial wastes has been analysed and the results are presented and discussed.

Algae from environments polluted with electroplating wastes have been reported by Ramaswamy and Somasekar[8]. They have studies the ecology of algae from streams carrying the effluents of Electro plating factory near Mysore City. The pH of the spots varies from 7.8 - 8.3. The algae found to occur in these environment as reported by Ramaswamy and Somasekar[8] are *Chlorella, Chlorococcum, Scenedesmus, Microspora quadrata, Cladophora glomerata, Pithophora, Cosmerium, Closterium, Synedra* and N*itzschia.*

3.1 Materials and Methods

The soil samples along which the electro plating wastes and coir waste flow were collected with clean spatula and brought to the laboratory. They were placed in sterile culture tubes containing sterile Chu 10 medium (20 ml) and incubated in light (Continuous illumination). They were allowed to remain under such conditions for one month and the algae in them were identified with the help of standard manuals[1,9]. The pH of the soil samples was also tested.

3.2 Results and Discussion

The soil receiving coir waste had a pH of 7.6. Under incubation, it showed the presence of the following algal taxa : *Nostoc, calcicola,* and *Phormidium, molle.* The soil receiving electro plate waste showed a pH of 8.1 and the following algal taxa : *Pinnularia fasciata, Nitzschia palea, Achnanthes hauckiana, Gomphonema parvulum, Oscillatoria obscura* and *Phormidium molle.* The pH of this environment is more or less the same as reported by Ramaswamy and Somasekar[8], but the algal constituents are different.

4 Case Study III : Brackish Water Pools

Some brackish water pools from Kerala have been analysed by Sankaran[10] for algal composition. The results are presented and discussed in this case study. Thalassery (Originally Telicherry) is a town along the Western Coast in Kerala. There are many backwater systems adjoining the seashore, on eitherside of the Eranjoli and Dharmadam rivers. Both these rivers merge with the sea in the Thalassery region itself. The two rivers originate from Wyanad area mountainous region, representing the Western Ghats. There are number of bracksih water pools on eitherside of the rivers. These are situated in the intertidal region. During monsoon, the shallow regions of these ecosystems are utilized for rice cultivation. This may be for a period of about 3 months. The average depth of the pond is about 4-5 feet. Prawn fishing is also done in this area throughtout the year. Coconut husks are soaked in these pools for defibering, by the cottage coir industries. These pools contain algal forms like *Bryopsis, Rhizosolenium, Spirogyra, Oedogonium* and diatom genera *Navicula, Pleurosigma* and *Gyrosigma* all of which have been reported from marine and brackish environments.

4.1 Material and Methods

Samples were collected from this brackish pool and the algae were identified using standard manual[1,11].

4.2 Results and Discussion

The following members of blue-greens were collected from these pools and they were described.

1. *Anabaena orientails*
2. *Anabaena sphaerica*
3. *Cylindrospermum doryphorum*
4. *Microcoleus chthonoplastes*
5. *Nostoc paludosum*
6. *Nostoc punctiforme*
7. *Oscillatoria princeps*
8. *Oscillatoria tenuis*
9. *Placoma vesiculosa*
10. *Scytonema sp*
11. *Spirulina major*

These brackish water pools appear to represent unique ecosystem containing certain algae from fresh water ecosystems. Many of the blue greens reported here, particularly of the genera *Anabaena, Cylindrospermum, Nostoc, Scytonema* are heterocystous and hence potential Nitrogen fixers. It is desirable to culture these organisms and estimate their nitrogen fixing potentialities. This will give a better idea to assess the quantum of contribution especially towards the nitrogen budget of these unique ecosystems. The presence of certain blue-green algae, particularly *Oscillatoria princeps* recorded here definitely indicates a trend towards eutrophication of the brackish water ecosystem.

5 Case Study IV : Aliyar River

Sankaran[12] has analysed the algal composition of the Aliyar riverine system and has compared the same within a time interval of 6 years. The results are presented in this case study.

The role of algae as bioindicators of river pollution will be considered here. Sometimes, pollution is difficult to detect chemically if the effluent passes off rapidly. But, biological analysis will indicate the extent to which the algal flora has suffered. The algae differ widely in their ability to withstand the various pollutants. Hence the degree of pollution can be assessed by estimating the algal communities and using algal indices at generic as well as specific levels. These indices will enable the worker to assess the degree of pollution. Nature is very unpredictable. A river may appear very clear, but it may be undergoing incipient pollution. On the contrary a highly polluted river may tend towards oligotrophic conditions by autopurification[13,14]. Both these conditions are well indicated by algal analysis.

Nygaard[15] proposed five indices to assess organic pollution of an aquatic ecosystem in terms of algal groups (myxo, chloro, eugleno phycean, diatom and the compound indices). The ability of the myxophyceae (Cyanophyta), euglenophyceae, centric diatom and chlorococcalean taxa to tolerate eutrophic conditions, and the common occurrence of pennate diatoms and the desmids under eutrophic conditions constitute the basis for the formulation of these indices. These indices are tabulated in Table 4. By calculating the values as pointed out, the river can be assessed as to the degree of eutrophication (pollution). These indices can be used with particular reference to organic pollution. Palmer[16] has made out a list of genera and species of algae tolerant to organic pollution. A pollution index of algal genera is also available. The level of pollution can be assessed by estimating the algal flora and adding up the index numbers.

Table 4 : Showing the Application of Nygaard's Trophic level Index to two sites of the Aliyar River 1984 and Current Algal Compositions.

Index	ANAMALAI		AMBARAMPALAYAM	
	1984	Current	1984	Current
1. Cyanophyceae / Desmids	4	0	0	3.5
2. Chlorphyceae: *Chlorococcales* / Desmids	1	0	0	0
3. Diatoms: Centric / Pennate	0	0	0.03	0
4. Euglenophyceae: Euglenophyceae / Cyano + chlorococcales	0	0.2	0	0.3
5. Compound: Cyano + Chloro coccales + centric diatoms + Euglenophyceae / desmids	5	0	0	4.5

Note : Criteria For Oligo/Eutrophic States

Index 1.	0-0.4	0.4-3.0
Index 2.	0-0.7	0.8-9.0
Index 3.	0-0.3	0.3-1.75
Index 4.	0-0.2	0.2-1.0
Index 5.	0.01-1.0	1.0-2.5

The total value so obtained will indicate the degree of organic pollution. The degree of pollution between different sampling spots can be easily compared.

The present work concerns itself with the application of these algal indices to the algal components of the Aliyar river along two spots. Comparisons are made between the current values and those obtained from algal composition of 1984[17,18].

5.1 Material and Methods

The river Aliyar has been studied with respect to the physicochemical characteristics and algal composition and succession[17]. Based on this investigation, ecology of this river has also reviewed[18]. In order to investigate the change that might have taken place with particular reference to the algal composition, a cursery algal analysis was undertaken. Water samples were collected at the same spots at Anamalai and Ambarampalayam along the Aliyar River, using standard procedures[15]. Samples were brought to the laboratory, filtered and concentrated. The algae were identified using standard manuals[17]. This is considered as the current year's collection. This collection was intensive and extensive for the two spots under consideration. But it does not cover the entire season. The list of algae occurring in both the spots was made out after identifying them upto the species level wherever possible. (Tables 2 and 3). Using Nygaard's trophic state indices and Palmer's pollution index of algal genera[15], the current algal composition was compared with the 1984 composition for Anamalai and Ambarampalayam (Tables 4 and 5).

5.2 Results and Discussion

Using the Nygaard's index, the algal composition of 1984 at Anamalai shows a eutrophic state (Table 4 : index 1 = 4 : index 5 = 5). The current index was `O' in both cases. This is because of the absence of desmids in the current collections. Based on the index, Anamalai is at the oligotrophic state at present, from what it was in 1984. On the contrary at Ambarampalayam, Nygaard's value indicates oligotrophic state in 1984, but eutrophic state at present. (Table 3 : Index 1 = 3.5 : index 5 = 4.5). Based on Palmer's index Anamalai was at 14 and Ambarampalayam at 19 in 1984. In the current composition the index value has increased to 24 for both the sites.

Nygaard's index seems to be of limited application as it is controlled more by a desmid flora. Based on this index it is to be concluded that the Anamalai tends towards oligotrophy. As revealed by Tables 1 and 2 in both the sites, along the Aliyar river there is change in the algal composition. It is to be particularly noted that genera *Chlamydomonas* and *Euglena* in Anamalai and *Chlamydomonas, Euglena, Lepocinclis* and *Stigeoclonium* in Ambarampalayam - all algal genera with high pollution index values have made their appearance in the current collections. All the genera mentioned above were altogether absent in 1984. It is possible that there is a very slight tendency for self purification at Anamalai, that too based on Nygaard's index values. It is to be further noted that there are abundant rotifer populations in the current collections of both the sites. Besides, there also pollution tolerant species of Palmer in the 1984 and current collections of both the sites. (These are indicated with * in Tables 2 and 3). Hence based on Palmer's index, occurrence of pollution tolerant species and occurrence of rotifers in the current collections, it is to be concluded that the Aliyar river is becoming more and more eutrophic. There are no industries at the sites in consideration. Human activities like bathing, washing of cattle and vehicles have brought about this process of eutrophication. Even a cursery reinvestigation points out to a possible eutrophication trend of this lotic system. Aliyar river, is the source of drinking water to this locality. The deterioration of its water quality must be prevented at all costs. Educating the public about the hazards of pollution and creating an awareness in their minds about the harmful changes in their source of water supply is of utmost importance. This can be achieved by voluntary organisation like Lion's, Rotary clubs, and NSS wings of educational institutions. Active participation by students, interaction of scientists with public, and drive from authorities will most certainly enable one to prevent the Aliyar river from becoming more eutrophic. Periodic monitoring of the river through water analysis and water assessment through biological algal analysis will be most highly desirable.

Table 2 : Showing the algal taxa present in the 1984 and in the current collections along the Aliyar river at Anamalai.

1984 Collections	Current Collections
CYANOPHYTA	**CYANOPHYTA**
1. Oscillatoria mougeotii	*1. Microcystis aeruginosa* *
2. Plectonema wollei	*2. Microcoleus paludosus*
3. Phormidium molle	*3. Oscillatoria annae*
4. P. rotheanum	*4. O. obscura*
	5. O. Subbrevis
	6. Phormidium molle
	7. P. anomala
BACILLARIOPHYTA- PENNATE	**BACILLARIOPHYTA -PENNATE**
1. Amphora coffeaformis	*8. Amphora coffeaformis*
2 . Achnanthes brevipes	*9. Achnanthes hauckiana*
3. Amphiprora Sp	*10. Gyrosigma sp.*
4. Cymbella cistula	*11. Gomphonema lanceolatum*
5. Cocconeis placentulua *	*12. Hantzschia amphioxys*
6. Caloneis schumaniana	*13. Navicula laterostrata*
7. Fragilaria intermedia	*14. Nitzschia closterium*
8. Fragilaria brevistriata	*15. N. palea* *
9. Fragilaria sp	*16. Pleurosigma sp.*
10. Gyrosigma scalproides	*17. Synedra ulna* *
11. G.distortum	**CHLOROPHYTA**
12. Gyrosigma sp	*18. Chlamydomonas sp.* *
13. Gomphonema parvulum *	*19. Spirogyra sp.*
14. Hantzschia amphioxys	*20. Ulothrix tenuissima*
15. Navicula cincta	*21. Oedogonium sp*
16. N. cuspidata *	**EUGLENOPHTA**
17. N. perigrina	*22. Euglena sp.*
18. N. halophila	
19. Navicula sp	Also recorded Rotifers *
20. Nitzschia obtusa	
21. N. vitrea	
22. Nitzschia sp.	
23. Pinnularia borealis	
24. P. fascinata	
25. Pleurosigma salinarum	
26. Pleurosigma sp.	
27. Rhopalodia gibba	
28. Surirella tenera	

29. Synedra ulna * *30. Synedra sp.*	
BACILLARIOPHYTA CENTRIC- NIL **CHLOROPHYTA** *1. Chlorococcum sp.* *2. Cosmarium divergens* *3. Rhizoclonium hieroglyphicum* *4. Spirogyra sp.* *5. Oedogonium sp.* **EUGLENOPHYTA - NIL**	 * *Pollution tolerant species*

Table 3 : Algal taxa present in the 1984 and in the current collections along the Aliyar river at Ambarampalayam

1984 Collections	Current Collections
CYANOPHYTA *1. Aphanocapsa roseana* *2. Oscillatoria sp.* *3. Microcystis aeruginosa* * *4. Nostoc punctiforme* *5. Plectonema wollei*	**CYANOPHYTA** *1. Oscillatoria princeps* * *2. O. obscura* *3. O. subbrevis* *4. O. chlorina* * *5. Nostoc piscinale* *6. Phormidium molle* *7. Rivularia sp.*
BACILLARIOPHYTA- PENNATE *6. Amphora coffeaformis* *7 . Achnanthes brevipes* *8. Amphiprora sp* *9. Cymbella cistula* *10. Cocconeis placentula* * *11. Fragilaria intermedia* *12. Fragilaria sp.* *13. Gyrosigma scalproides* *14. Gomphonema parvulum* * *15. G. lanceolatum* *16 Hantzschia amphioxys* *17. Navicula cincta* *18 N. cuspidata* * *19. N. perigrina* *20. N. halophila* *21. N. Salinarum* *22. Navicula sp*	**BACILLARIOPHYTA- PENNATE** *8. Amphora coffeaformis* *9. Epithemia zebra* *10. Fragilaria intermedia* *11. Gyrosigma sp.* *12 Gomphonema parvulum* * *13. Navicula salinarum* *14. Pinnularia fasciata* *15. P. interrupta* *16. Pleurosigma sp.* *17. Rhopalodia gibba* *18. Synedra ulna* * **CHLOROPHYTA** *19. Chlamydomonas sp.* * *20. Cosmarium sp.* *21. Closterium* * *22. Stigeoclonium sp.* * *23. Spirogyra sp.*

23. *Nitzschia obtusa* 24. *N. closterium* 25. *N. Palea* * 26. *Nitzschia sp.* 27. *Pinnularia borealis* 28. *P. interrupta* 29. *Pleurosigma salinarum* 30. *P. angulatum* 31. *Pleurosigma sp.* 32. *Rhopalodia gibba* 33. *Surirella tenera* 34. *Synedra ulna* * 35. *Synedra sp.* 36. *Tabellaria intermedia* 37. *Tabellaria sp.*	24. *Oedogonium sp.* 25. *Euglena sp.* 26. *Lepocinclis sp.* * Also Recorded Rotifers *
BACILLARIOPHYTA - CENTRIC 38. *Cyclotella meneghenia na* * **CHLOROPHYTA** 39. *Oedogonium sp.* 40. *Rhizoclonium hieroglyphicum* 41. *Scendesmus quadricauda* * 42. *Spirogyra sp.* **EUGLENOPHYTA - NIL**	* *Pollution tolerant species*

In conclusion, it may be emphatically stated that edaphic and aquatic systems are damaged even by cottage industries that use soil and water bodies in a manner that is far from judicious. Industries such as dehusking units are adequate to damage an ecosystem. That the damaged ecosystem are not beyond repair is underlined by the presence of not only pollution indicator species of algae but also of many beneficial blue-green algae borne out of the investigative analyses of cement dust polluted soil, coconut husk loaded soil and brackish water ecosystems. The algal communities, particularly the blue-greens are tolerant to extremely adverse conditions of the environment by virtue of which ability they serve as pollution tolerators and indicators.

The genera of blue - greens *Phormidium, Oscillatoria, Microcoleus* and *Nostoc* are common to all the case studies taken and considered here. The algal taxa in general and the blue greens in particular are not only important as primary producers but also tolerators and indicators of pollution in the damaged ecosystem.

Table 5 : Showing the Application of Palmer's Pollution Index of Algal Genera to two sites of the Aliyar River 1984 and Current Algal Compositions :

Algal Genera	ANAIMALAI		AMBARAMPALAYAM	
	1984	Current	1984	Current
Chalamydomonas	-	4	-	4
Closterium	-	-	-	1
Cyclotella	-	-	1	-
Euglena	-	5	-	5
Gomphonema	1	1	1	1
Lepocinclis	-	-	-	1
Microcystis	-	1	1	-
Navicula	3	3	3	3
Nitzschia	3	3	3	-
Oscillatoria	4	4	4	4
Phormidium	1	1	-	1
Scenedesmus	-	-	4	-
Stigeoclonium	-	-	-	2
Synedra	2	2	2	2
Total Index value	14	24	19	24

Number = Pollution Index value, -absent

Acknowledgement

The author wishes to express his thanks to the Principal and Management of the N.G.M College Pollachi for facilities.

References

1. V. Sankaran, *Indian J. Environ. Hlth* **19**, 67 (1977).
2. T. V. Desikachary, *Cyanophyta*, I.C.A.R, New Delhi (1959).
3. S. Parthasarathy, N. Arunachalam, K. Natarajan, Obilisamy and G. Rangasami, *Indian J. Envirion. Hlth.* **17**, 114 (1975).
4. G. E. Fogg, *Symp. Soc. exp. Biol.* **23**, 123 (1969).
5. R. E. Cameron, *Trans. Am. Microsc. soc.* **81** , 379 (1962).

6. R. N. Singh, *Role of Blue - green algae in N_2 Economy of Indian Agriculture,* I.C.A.R., New Delhi, (1961).
7. V. Sankaran, M. Sambasivamoorthy and I. Santhanaraj, *All India Symposium on Algae and Human Affairs and Environment,* Pushpam College, Poondi, India (1991).
8. S. N. Ramaswamy and R. K. Somasekar, *Phykos* **21**, 83 (1982).
9. G. Venkataraman, Systematic account of some South Indian diatoms. *Proc. Indian Acad. Sci.* **(B)** **10**, 293 (1939).
10. V. Sankaran, *Sea Weed Res utiln,* **15**, 205 (1991).
11. K. Biswas, Records of the Botanical survey of India, **XV** (1), Common Fresh and brackish water algal flora of India and Burma (reprinted) Periodical Experts Book Agency, Delhi (1980).
12. V. Sankaran, *National symposium on Biomonitoring indicators in an aquatic ecosystem,* Vellalar College for women, Erode (1990).
13. A. Mahadevan and Krishnaswamy , *Polln. Res.* **5**, 69 (1969)
14. A. Ranjithakani. *Papers of FW. Biology - Memoirs of Salem Institute of Experimental Biology,* **No.1**, 222 (1981).
15. R. K. Trivedi, and P. K. Goel, *Chemical and Biological methods of water Pollution studies* - Enviro Media Publications, Karad (India), (1984).
16. C. M. Palmer, *J. Phycol.* **5**, 78 (1969).
17. V. Sankaran, *Phykos,* **23**, 75 (1984).
18. Sankaran, in *Ecology and pollution of Indian rivers,* ed. R.K. Trivedi (1988).

THE MANGROVES OF KERALA (INDIA): STATUS AND RESTORATION

C.M. JOY

Sacred Heart College, Cochin 682013 Kerala, India

AMMINI JOSEPH

School of Environmental Studies, Cochin University of Science and Technology, Cochin 682016 Kerala, India

1 Introduction

Mangroves are the salt-tolerant ecosystem found in the tropical and sub-tropical intertidal regions comprising of a unique association of plants and animals. The mangrove ecosystem tends to be very productive and is the source of timber as well as many commercially valuable products. The role of the mangrove vegetation in coastal soil binding and as a filter bed of pollutants has been widely recognized.

The total area of mangroves in India is estimated to be 6740 sq.km. which is about 7% of the world's mangroves. Out of this the Sunderbans of West Bengal has the largest area followed by the Andaman and Nicobar Islands. The remaining are scattered in the states of Andhra Pradesh, Tamil Nadu, Orissa, Maharashtra, Goa, Karnataka, Lakshadweep and Kerala[1].

History records that the coastline of Kerala ramified by backwaters and shallow canals were once fringed by luxuriant growth of mangroves. About 70,000 hectares of mangrove land existed in Kerala which has now got reduced to a few isolated patches consisting of a few species[2]. It is estimated that about 44% of the mangroves were destroyed during 1975-90. The important patches of mangroves existing now in Kerala are distributed in Chiteri, Kungimangalam, Pappinisseri, Edakkad, Thalassery, Nadakkavu, Chetuvai, Kannamali, Kumarakam, Quilon and Veli (Figure 1). Perhaps the longest chain of natural mangroves vegetation along the Kerala coast may be at Kumarakam on the banks of Vembanad estuary[3].

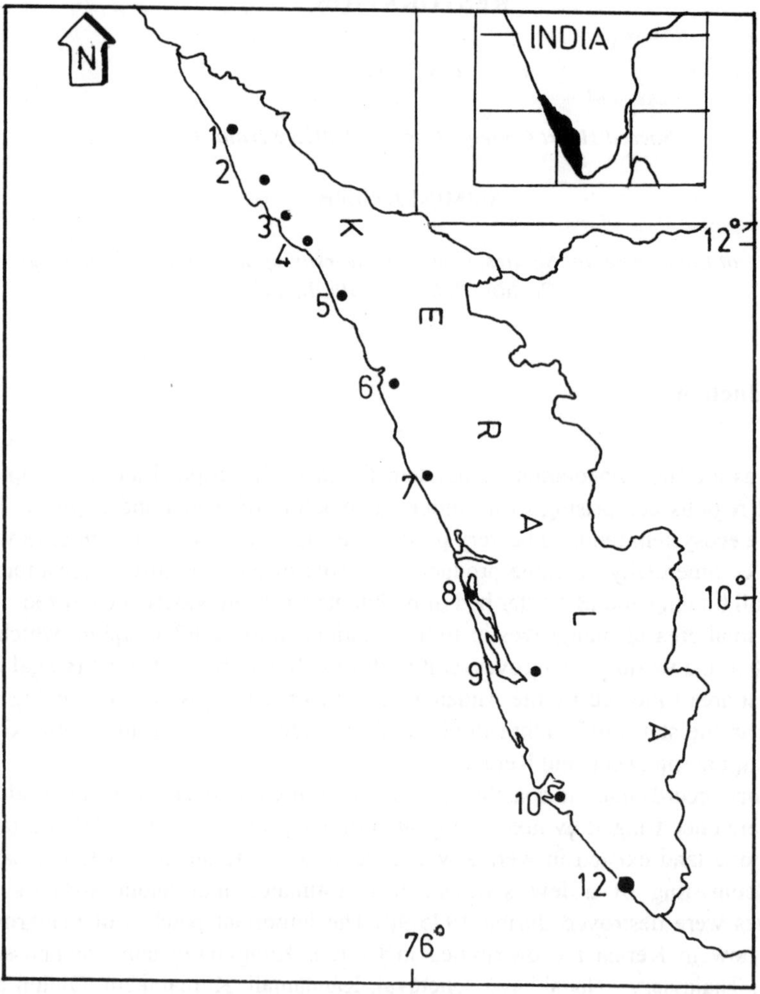

1.	Chiteri	2.	Kungimangalam	3.	Pappinisseri
4.	Edakkad	5.	Thalassery	6.	Nadakkavu
7.	Chetuvai	8.	Kannamali	9.	Kumarakom
10.	Quilon	11.	Veli	O	Thiruvananthapuram

Figure 1 : Locations of important mangrove vegetation in Kerala

2 Mangrove Flora

One of the earliest references on the mangrove flora of Kerala is obtained from *Horthus Malabaricus Indicus* which cites eight species. Bordillon[4] reported *Brugiera gymnorhiza* and species of *Rhizophora* as being very common in regions of Quilon; but Erlanson[5] has established the disappearance of these species along this coast. He has also reported the deforestation of mangroves for the construction of the Cochin Port. Rao and Sasthry[6] reported the occurrence of *Avicennia officinalis, Rhizophora mucronata, R.. apiculate, Brugiera gymnorhiza, Sonneratia, Cerbera manghas, Derris trifoliata, D.heterophylla* and *Acanthus ilicifolius* at Veli, a small meandering coastal lagoon joining the sea. The occurrence of *A.. ilicifolius, Avicennia alba, Rhizophora sp.* and *Brugiera sp.* in small numbers was reported by Kurian[7].

Suma[8] has studied the distribution of mangroves in Vypeen region of Cochin. She reports eight true mangrove species; they are *Rhizophora mucronata, Brugiera cylindrica, Acanthus ilicifolius, Avicennia officinalis, Excoecaria agaelocha, Brugiera gymnorhiza, Sonneratia caseolaris* and *Rhizophora candelaris.* The mangrove associated flora consist of mainly *Clerodendron inermi, Derris ulignosa, Cuscuta reflexa* and *Cerbera odollam.* The less frequent species are *Calophyllum inophyllum, Polygonum glabrum, P.barbatum, Sphaeranthus indicus, Calonyction boxa-nox, Ipomaea pes-caprae, Hygrophila anguistifolia, Moniera cuneifolia, Aeschynomene indica, Erythrina indica, Crotalaria striata, Thespesia populnea and Casuarina equisetifolia.*

The other species of mangroves reported from Kerala are *Avicennia caseolaris, A.. marina* and *Candelia candal.* The associated flora found along the coast also include *Manileara hydroza, Phoenix humilis, Dalbergia candinattus, Calamus rotang, Tylophora tenucima, Lumnitzera recemosa* and *Viscum oriental*[9].

The mangroves on the shores of Cochin backwater are mostly formative developing on small reclaimed or natural islands[10]. A common observation is that the reclaimed lands are first colonized by *Avicennia* and halophytic grasses followed by *Rhizophora* and *Acanthus ilicifolius*. As the soil building proceeds other mangrove associated species appear. These habitats are highly vulnerable to human encroachment.

3 Restoration

In spite of the favourable climate there are no larger formations of mangroves in Kerala. The most important reason for this may be the population pressure. Reclamation of the estuarine coasts for human settlements has been in progress since the 1970s as a result of which mangroves were extensively destroyed. Industrialization and aquaculture activities may be a secondary cause for the disappearance of mangrove flora. Added to this is the public feeling that wetlands are wastelands as a consequence of which they were used as dump-yards of wastes. According to Thomas and Fernandez[11] soil dredging has also contributed to the destruction of mangroves. The net result is that the mangroves are reduced to a few patches that is mostly under private control.

With the rise of environmental movements many suggestions have been put forward and implemented to conserve the ecosystem. The concept of mangrove tourist resorts is gaining ground. Mangrove locations such as Veli, Kumarakom and Neendakara are already well known tourist centres. It is suggested[12] that strong directives from the government is needed to protect these spots as natures reserves. Some of the important suggestions in this regard are, as the mangrove areas are converted to tourist resorts the administration can recruit officials to protect and monitor the mangrove species of flora and fauna by fresh plantation and zealously protecting those that are surviving. This can be done by ear-marking core areas not be trespassed by tourists and fringe areas where tourists are permitted[11]. The "Mangalavanam" at Cochin peculiar for its mangrove avian fauna is under severe stress due to its location in the midst of a port city. Plans have been drawn up by the civil authorities to restore it as a mangrove tourist spot within the city. Further south, the Pathiramanal Island in the Vembanad estuary fringed by mangroves has also been ear-marked as a mangrove tourist centre.

The State Committee on Science and Environment has supported the planting of mangrove seedlings in Veli estuary successfully. The environmental NGOs in Kerala have done some isolated attempts to protect the mangroves. This include planting of the seedlings of *Rhizophora* sp. in the edges of the estuarine channels in central and northern Kerala. Except for this there has been no concerted attempts to plantation of mangroves in the state. It is suggested that an extensive awareness programme related to the value of mangroves, conservation and utilisation among decision makers and the public should be urgently instituted. This is possible by organizing a mangrove development camp involving the local school and college students as well as teachers to make them involved in the process of coastal protection. Of equal importance is that the government should have a policy that includes planting of mangrove trees incorporated in the coastal afforestation programmes.

References

1. P. D. Sharma, *Ecology and Environment.* Rastogi Publ. New Delhi (1992).
2. S. C. Basha, *Indian Forester,* **120**, 175 (1992).

3. K. K. Ramachandran, and C. N. Mohanan, *Proc.Natn.Sem.Estuarine Management* Thiruvanathapuram (1987).
4. T. F. Bordillon, *The Forest trees of Travancore.* Trivandrum Govt. Press. Trivandrum (1908).
5. E. W. Erlanson, *J.Ind.Bot.Soci.* **15**, 193 (1936).
6. T. A. Rao, and A. R. K. Sasthry, *Indian Forester,* **100**, 438 (1974).
7. C. V. Kurian, *Asian Sym. Mangrove Environ.Res.Mangams.* University of Malaya, Kaulalampur (1980).
8. K. P. Suma, M. Phil. Dissertation Thesis. Kerala University, Thiruvananthapuram (1995).
9. STEC, Science, Technology and Environment Committee. Thiruvananthapuram (1995).
10. M. S. Rajagopalan, C. P. Gopinathan, C. S. G. Pillai, P. P. Pillai, D. Selvaraj, P. M. Aboobacker and Kanakam, *Proc. Symp. Coastal Aquaculture.* Cochin (1986).
11. G. Thomas, and T. V. Fernandez, *Indian Forester,* **120**, 406 (1994).
12. A. Mohandas, Ph.D. Thesis. University of Kerala (1993).

INDIAN ETHOS FOR RESTORATION OF ECOSYSTEM

A. S. REDDY

Department of Biosciences, Sardar Patel University
Vallabh Vidyanagar - 388120 India

1 Introduction

Despite the vastness of the Universe, life exists only in a limited thin enveloping layer of a single planet, 'the earth'. Within this life suppoting layer, the biosphere, life exists in a great varied forms. We precisely call this wide range of living forms as Biological Diversity which encompasses the totality of variability at gene level, species level and ecosystem level. The richness of this biodiversity is governed by various complex biotic and abiotic components of the environment including us, the human beings. Though we belong to only one species out of many million of species, the lifeforms of the biosphere are much influenced by this single species - *Homo sapiens* due to its dominated over growth, over consumption and over (so called developmental)- activities. As per the basic ecological principle, whatever extracted over a period from the ecosystem should be equal to the effective regeneration of the same during an equal period. Then only the ecosystem sustains for a long time. We, the human beings today, unlike the other biotic compounds of the ecosystem, however, consume the more share of resources from the nature and in reply give back more waste, often non degradable and hazardous to the nature. In our path of building the modern civilisation, the ecological balance was disturbed, societies were plundered, cultures were destroyed and a turmoil was created in the present system. Thus the indiscriminate activities of human being not only caused severe damage to the ecosystem in which he lives but also lead his life into peril. Realising this endangered situation of his existence on this planet, the modern man recently started to make efforts for repairing the damaged ecosystem and restoring it to the nearest possible level of naturality. At this juncture it may be quite reasonable to look back at the history of our development and get the possible answers to the question, how our ancestors lived in harmony with nature for centuries. The most environmentally polluted western societies also started keenly observing the merits of the cultural heritage of Indian Society where the ecological ethics and traditions maintained the balance between nature and the behaviour of common Indian. But the Indian Society is in dilemma whether to mimic the western fast development, shedding its time tested culture or to adhere to its own traditional path towards slow but sustainable development. The present documentation is an attempt to highlight the relevance of Indian ethics, traditions, customs in the context of restoring the damaged ecosystems.

Cultural Diversity of India

The cultural diversity which exists in the Indian society reflects the intimate relationship between the existence of humanlife and all other living and nonliving creatures in nature. Conservation of nature, natural resources and magnitude of the biological diversity etc.; have been deeply rooted in Indian traditions and culture. The concept of care for nature and its creatures goes back to the ancient time. The treaties and literature such as the Vedas, Puranas, Upanishads of ancient India depict clearly the relationship between man and the nature.[1] They state that since time immemorial, Indians have known that all life forms depend on nature. We have to be very clear that by nature, we do not mean just the greenery around us. It is the matter of attitude to the principles of laws prevalent in the cosmos and the way we apply these principles in our day-to-day life. The day-to-day activities, rituals and rites of a common Indian are, even today, in harmony with natural laws and principles. The houses in rural India or the buildings in towns and cities of ancient India, unless there is no domination of westernisation, are built according to the direction of the sunlight and wind. Only in recent times because of the over concentrated population in towns and cities by the process of urbanisation, modernisation and westernisation, the constructions are planned in such a manner that we are gradually moving away from nature. Unlike in the past, today we are forced to depend upon artificial illumination in the buildings rather than the natural sunlight. Otherwise the common Indian starts feeling uneasy, the moment he is removed from his natural environment. He never had merely an utilitarian relationship with nature. For him, nature is not just a conglomeration of objects, but it is a complete system. He is well aware of the fact that we are a part of nature but not apart from the nature nor are we over and above it. We can not conquer the nature. The Iso-upanishad thus explains that *"The Universe along with its creatures belong to the Lord. No creation is superior to any other. Human beings should not be above nature. Let no one species encroaches into the rights and privileges of other species."*[1] Indeed the present day's advanced materialistic infrastructure can not reflect the totality of Indian's understanding of the deep and essential relationship between man and every other component of nature. We find nature being kept as focal point in all the physical and spiritual activities. Because of such a deep commitment towards the nature, Indians were never allowed themselves to be far removed from the nature nor did they differ from natural laws while making his path towards development. They always have had reverence to all the creatures of nature. Indian culture rightly recognised the significance of the five basic elements of nature, the earth; the water; the air; the light and the cosmos. Indians consider that these five elements, the *'Panchbhutas'*, are the five forms of divine and thus worship them in reality and symbolically in various forms. Indeed all elements of nature such as mountains, lakes, rivers, plants and animals have been perceived by Indians as deviant source.

From the point of regulation of any ecosystem, we know very well the significance of solar energy. The daily rituals of a common Indian, such as Surya Namskar, i.e., worshipping the Sun implies his awareness of the ultimate source of energy and the reverence for this endless natural resource. According to Yajur Veda, the whole universe is full of energy in which the Sun is at the centre and the ultimate source of energy for all living organisms on earth. The net energy flows from the point of production to the point of consumption through the plants, animals, human beings. Vegetarianism, which is not just a form of Indian diet but the lifestyle, also supports the most effective use of Sun's energy. Because, the food chain of vegetarians is the shortest and a given area of land can support large number of people. We know that one acre of land grows 10,000 kg of potato which can support many people; the same land at the most can grow enough grass or fodder to produce only 80 kg of meat of an herbivore. It requires 250 litres of water to grow grass and fodder to produce one kilogram of meat. Thus the food chain of Indian vegetarianism minimises the transactional loss of energy flow. These ecological principles are well pronounced in Rig Veda. It says (R.V.8.20.25) "One should take from rivers, mines, mountains and seas what is useful and according to one's need and without injuring them...."

In India, all the rivers such as Ganga, Yamuna, Saraswati, Narmada, Kaveri, Krishna, Godavari, etc., are equated with Goddesses and are worshipped. Traditionally, Indians add the ashes of dead bodies into these sacred rivers; considering the merit of calcium and other purifiers present in the ashes. Similarly in India it is a common practice, of offering coins, especially which are made of copper, into the rivers. This concept has the hidden fact of purifying the river waters[2]. However, due to cultural and ecological negligence in modern times, in the name of ashes of 'Haven' half cremated dead bodies are being thrown into sacred waters. Unfortunately owing to large scale industrialisation and devouring consumerism we are loosing that identification with nature.

The Indian festival of cow Nourisher worship[3] which was explained by a story, in the Sanskrit literature of 20th Century A.D., the Bhagavata Purana clearly indicates the recognition about the significance of the living ecosystem as a whole. Though the story concerns about Lord Krishna's adventures during his childhood with his cowherd companions at a hill named Govardhan (Gobardhan), it has got a great ecological implication. The meaning of the Sanskrit name Govardhan (Go=Cow; Vardhan=nourishing) is cow nourisher. In some parts of Eastern India, the same word is pronounced as 'Gobardhan' instead of Govardhan. Krishna directs the Cowherds to worship Govardhan hill, the natural resource which support the livelihood of their cattle. He condemns worshipping the distant unseen gods such as Indra, the God of Rain. Accordingly they start worshipping the hill rather than God. Then Indra gets angry over the defection of his worshippers and creates violent rain storms to destroy the cowherds and their cattle. Krishna then lifts the Goverthan hill on his little finger and provides them shelter and ultimately saves all the cowherds and their animals. Every year at the actual hill of Govardhan in Mathura District of Uttar Pradesh the festival is celebrated to worship the hill which is a complex unit of life supporting system.

The little modified version of this story reveals us another ecological implication. The villagers of Kishan Garhi in Aligarh District of Uttar Pradesh celebrate this festival with a different meaning. Instead of accepting the meaning Cow Nourisher (Go + Vardhan), they consider it as Dung Wealth (Gobar = dung; Dhan = Wealth). Hence during this festival the women and children of the village make a model of small hill with Cowdung, representing the sacred hill. With straw and cotton, trees are represented on the model. Little models of cowboy and cowherd are also made of dung to represent probably Lord Krishna and the cattle. They place a lamp on the model, wind thread around the tree models on it. This part of culture is also represented in many parts of India. Keeping aside the religious or cultural aspect in either of these two events, it is a matter of appreciation from the ecological point of view that Indians had reverence not only to the living components but also to the abiotic components of the ecosystem. In contemporary Indian society, the dung has been variously valued as of 'National Wealth', 'diamond' and so on. In Indian traditional agricultural system, cattle dung has remained as one of the most important and basic inputs till it has been replaced during the recent agricultural revolutions with chemical fertilizers. Within a short span, this eroded traditional agricultural practice resulted the great amount of degraded soils and thereby we are switching once again to the inputs of farm yard manure.

Sacred Concept : A means of Conservation

As far as the conservation of plant and animal diversity is concerned Indians have rich cultural heritage which considers all kinds of plants and animals as 'Sacred'. Animals and other creatures which faced a danger from man, were termed sacred by associating them with some religious rituals so that people could live in harmony with them.[4] The concept of Dashavatars, the ten forms of incarnation and phases of the divine, not only reflects our reverence to the broad spectrum of animal kingdom but also reminds us of the various stages of evolution of life from fish (Matsyavataar = aquatic form) to tortoise (Kurmavataar = amphibian form) to boar (Varahavataar = terrestrial form) and so on. Worshipping all these forms of life clearly mirrors our consciousness of the conservation of all kinds of animals, whether they are of wild or domesticated. Indians worship various birds like peacock (Mayur), eagle (Garud) and the animals such as rat (Mushika), ox (Vrishabha), tiger (Shardul), lion (Sinh) and so on. We worship all these living components of the ecosystem by assigning them as vehicles to Gods/Goddess : Karthikeya, Vishnu, Ganesh, Shiv and Durga respectively. Worshipping serpent god (Nagdev) and Cow goddess (Go-Maata) is a phenomenon very commonly observed throughout India. This is nothing but an expression of our reverence for all these animals from cultural, religious, economic and ecological points of view.

There exists lot of literature which emphasizes the protection of all kinds of animals. Providing food for creatures like insects, birds, and animals has been regarded as an extremely virtuous act. Just as it is customary to provide food to the hungry and poor, making similar provisions for other living creatures is also regarded as a virtuous deed. This concept has so deeply been rooted into Indian society that today it has become a custom for the Indian to offer a handful of his daily food placing it in the open courtyard for living creatures before he consumes it. While doing so, he believes that God accepts his food in the form of different animals or birds. There are innumerable stories which narrate the interest and morals linked with the protection of animals or birds. A verse from Yajur Veda (13.37) describes *"O king, you should never kill animals, like bullocks useful in agriculture or like cows which give us milk and all other helpful animals, and must punish those who kill or do harm to such animals."*

Narsimhapuran (13.44) emphasized prohibition of killing or eating birds by saying *" O, wicked men ! if you kill a bird then your bathing in a river, pilgrimage, worship and yagnas are all useless."* In Vishnu Puran (3.8.15) it is said " God Keshava is pleased with a person who does not harm or destroy other creatures or animals."

The following story from the ancient literature reveals the great concern of Indians about the protection of animals. A king named 'Shibi' was well known for keeping his promises and protecting all the living creatures of his kingdom. Once a dove with fear rushes to him while it was being chased by another larger carnivorous bird, and urges the king to save her life from the danger. The king immediately assures the dove, considering it was his duty. Soon after, the carnivore also comes to the king and requests him to handover the dove. When the king refuses, the predator alarms that the former couldn't stop the latter's prey as it was against the law of nature. Realising the critical situation, the king offers the flesh, of equal weight of the dove, from his body to the predator bird so as to save the dove's life as well as to satisfy the hunger of the carnivore bird.

Similarly Indian history also reveals about the concern of the Emperor Ashoka (273-236 B.C.) for preservation of wild animals and plants. Pillar edicts of Ashoka period depicts his great care towards the welfare of natural creatures, flora and fauna including the wild beasts like tigers and lions. He prescribed various financial punishments for killing animals even squirrels and ants. For the first time in world history a sanctuary for the preservation of wildlife was created by Ashoka. Wild animals like tigers, lions, leopards, bears and bisons were given shelter and protection. The following is the verse from pillar edict VII the Ashoka's period[5] :

The Beloved of the gods,

When I am crowned twenty six years, these various animals are declared by me inviolable, viz, parrot, mainas, the aruna, ruddy geese, wild ducks, the nandimukha, the gelata, bats, the amba-kapilika, small tortoises, boneless fish, the vedaveyaka, the Ganga-puputaka, the sankuja-fish, large tortoise and porcupines, squirrels, young deer, bulls, okapinda, wild asses, white pigeons, village pigeons and all Quadrapeds which neither useful nor edible.

Those she-goats, ewes and sows which are either young or are giving milk to their young are inviolable, and so also are those of their young ones which are less than six months old.

Cocks are not to be caponised.

Husks containing living beings (insects) are not be burnt.

Forests are not to be burnt, either uselessly or for killing animals.

One animal is not to be fed with another animal.

On the three Chaturmasis, on these three days during Tisya full-moon viz. the fourteenth, the fifteenth, and the first tithi- and invariably on every fast-day, fish are inviolable and not to be sold.

On these very same days, those other classes of animals (that live) in elephant parks and in fishermen's settlements, are also not to be slain.

On the eighth (tithi) of every (lunar) fortnight, on the fifteenth, on Tisya, on punarvasu, on the three chaturmasis and on auspicious days, bulls are not to be castrated and he goats, rams, boars and other animals that are usually castrated, are not to be castrated.

On Tisya, on punarvasu, on Chaturmasis and during the fortnight of every Chaturmasi, the branding of horses and bullocks is not to be done.

Till I had been crowned twenty six years - during this period, prisoners were released by me twenty five times.

Regarding the conservation of green wealth i.e., the spectrum of plant diversity is concerned; it is an integral part of Indian society since time immemorial. In Indian tradition the tree has been known by the name of 'Visvapurusa' i.e., the universal soul. Our relationship with trees are moral and ethical. In Sanskrit literature there are many stories in which an old man while planting a fruit bearing tree is asked "why are you planting this tree ? You are not left with many years to eat its fruits." The old man's answer invariably is that he has been enjoying the fruits of the labour of his ancestors, and therefore, he owes a duty to his future generations. He may not reap the fruits of this tree but his future generations would surely do so.* This simple story reflects ecological ethics for the sustainability of the system. Planting trees in one's surroundings and protecting vegetation with the purpose of providing shelter to birds and animals has been regarded, in India, not merely as one's duty but as an act of piety.

To the aborigines of Indian jungles, trees mean a good deal more. The words for tree and house are practically the same among some of the aboriginal tribes of India.[4] Ancient Indians never regarded the trees as a commercial commodity. To keep people's interest in trees alive, lot has been said in praise of trees in Indian literature, tables, traditions and customs. There exists a belief that planting a tree on the road side can be more worthy act than giving birth to unworthy children.

Varaha Purana conveys the similar meaning in the following :

अस्वथमेकम् पिकुमिन्दमेकम् न्याग्रोधमेकम् दश पुष्पजातिः
द्वै द्वै तथा दाडिमामातुलुंगे पंच आम्रोपि नरकम् न याति ॥

"Aswathamekam, Picumindmekam, nyagrodhamekam, das pushpajati;
Dwy-dwy tatha dadimamatulunge, panch amropi narakam na yati."

It conveys that one who plants one pipal (fig), one nima (neem) one vata (banyan), ten flowering plants or creepers, two pomegranate, two orange and five mango trees does not go to hell. From ecological point of view, this concept reflects the significance of planting various kinds of plants i.e., the polyculture. Different kinds of plants mentioned above are not only useful to human beings, but they also protect various animals, birds, insects etc., by providing them food and shelter. Thus, we can infer that the present day's global burning issue of biodiversity conservation has been well taken care of in ancient India. Monoculture in India is resulted only in recent times when the traditional society has been influenced by the commercialised modern society.

As the society passed on from the hunting stage to agricultural stage, the tribe continued to hold the ancestral trees in reverence. Though the forests were cleared for cultivation, considerable number and size of natural groves were invariably left in the clearings. These groves near villages of tribals, in course of time, have been familiarly recognised as 'sacred groves.' All forms of vegetation in such a sacred grove including shrubs and climbers are invariably protected. From the ecological point of view these groves are the reservoirs of biological diversity. They harbour the vegetation in its climax formation and also various animal forms. The plant and animal species are found to be safe in such groves which are otherwise rare and endangered in the forest. Removal of any living or dead material from such groves is a taboo. Such a taboo prevailing in a small village named. 'Gutribayalu' of Anantpur district in southern part of India protected a banyan tree for about 600 years and brought the tree into 'Guinness Book of World Records' as world's largest banyan tree.[6] The tree, because of the sacred belief, could successfully spread its crown canopy over an area of 5.2 acres with its 1500 prop roots and consequently broke the earlier world record made by the banyan tree of Botonical Gardens Calcutta. For three days following shivratri, every February, about 30,000 people gather at this tree to worship. This reveals us that the age old sacred concept, from the conservation point of view, is much more stronger than the recent scientific means.

Only on extraordinary occasions, such as a disaster in the village, the wood is collected from these sacred groves, that too, after performing a special ritual. Even today the common tribesman in India, by performing some rituals, formally seeks permission of the devine spirit of the tree before applying his axe to it, whether in a forest or at any other place. This practice has got the direct linkage with the rituals performed by the Indian farmer prior to harvesting the crop from his field. Before he starts consumption of such newly harvested crop, a small part is offered to animals, birds or even to human beings in the name of God. All these acts are in accordance with the ecological principles.

Tree worship is an integral part of Indian culture since time immemorial. It is said that tree worship was widely prevalent in India much before temples came into being. In Mahabharata, the Adiparva said :

ऐको वृक्षोहि यो ग्रामे भवेत पर्णफलान्वित:
चैत्योभवति निर्ज्ञातिरर्चनीय सुपूजिता:

eko vrkso hi yo grame bhavet parnaphalanvitah
chaityo bhavati nirjnatirarchaniyah supujitah

It means whenever a tree is found in full bloom covered with leaves, flowers and fruits in a village it becomes worth worshipping. In this regard Rig Ved stated *"If you want to enjoy the fruits and happiness of life for hundreds and thousands of years, take up systematic planting of trees.*

In India there has been a long tradition of planting trees and growing them by the side of wells and ponds situated in a central place or a public place. These trees are worshipped on all ceremonial occasions. The important ecological significance of worshipping these trees is that such places are being maintained well from time to time. Thus tree worship reflects the ecological awareness.

The well known concept of Pancavati (Panca = five; Vati = grove) which is a tradition pervading the whole of India tells us the deeper understanding of Indians about the significance of various kinds of vegetation and individual components of such vegetation. In a broader sense the term Pancavati means a forest with five kinds of trees. But the word 'five' has dual meaning. It conveys a sense of totality or plurality.[7] That is the reason why, various kinds of trees and plants are worshipped in different parts of India. Some trees may be common throughout the country. Vata (*Ficus benegalensis*), Peepal (*Ficus religiosa*), Asoka (*Saraca indica*), Bel (*Aegle marmelos*), Tulsi (*Ocimum sanctum*) are of this category. Whereas the Neem (*Azadirachta indica*), Shami (*Prosopis cineraria*), Coconut (*Cocos nucifera*), Aak (*Calotropis* sps.), Mango (*Mangifera inidca*), Asupalao (*Polyalthia longifolia*), Amla (*Phyllanthus emblica*) etc. are more or less regional specific for worshipping. The plants like basil (Tulsi) are worshipped by families individually, while some other plants like peepal are worshipped at public places. These are only few out of many to mention the names of plants which are worshipped in India. All these different kinds of plants are considered sacred and worshipped either directly or by using them for worshipping various Gods on different occasions. For instance, Indians do not even think of starting any good work either small or large without invoking Lord Ganesh, the God of knowledge. More than thirty kinds of plants are used in Ganesh puja. The ecological merit of this worship is that most of these plants are of wild category. Thus the practice of worshipping trees and plants has helped the Indian society in preserving the link with the world of plants and trees and with the nature.

We find that in India, there will be usually a neem tree after every three houses. This tree has been regarded as Goddess 'Lakshmi' who keeps us ever healthy and wealthy. The significance of this neem tree has been recognised by the scientific world only very recently. Similar is the case with our basil herb. It is almost essential to grow a basil plant in the courtyard and it is looked after the way a god is looked after. Its sanctity is strictly maintained.

In Sanskrit literature, on different occasions, it is said that a person lives in heaven (Vaikuntha) for as many thousand years as the number of days a basil plant blooms in his house. If one plants a Bel (*Aegle*) tree, prosperity reigns in his house for generations. One who plants a peepal tree he attains the abode of Lord Vishnu; while he attains Sivaloka, if planted a banyan tree. A person who plants neem trees attains Suryaloka for three epochs. One who plants mango trees either in a park or on roadside, emancipates fourteen ancestors or descendants and so on. Thus tree plantation and tree worship are deeply rooted in Indian culture. In contemporary India, since 1950, these programmes have been celebrated as National Festival of Trees or Van Mahotsava. Every year on this festival, lakhs of seedlings and saplings are planted all over the country with a reverence to the green trees.

Protection of Environment in Indian History

Destruction of vegetation and cutting of green trees have been considered to be sin and sacrilege. Acharya Charak, the originator of herbal medicine in India and the father of science of Ayurveda in his book 'Charak Samhita' around 5th century A.D. condemned the destruction of forests in the following words[8] :

विगुणेश् वपितु खलुएतेषु जनपदोध्वंस करेशुभावेषु
भेषजेनोपपाद्य मननम न भयम् भवति रोगभ्यस्ति।

Vigunesh vapitu khalueteshu janpadodhwansh kareshu bhaveshu
Bheshjenopapaday mananam na bhayam bhavati rogabhyasthi

It conveys that destruction of forests is most dangerous for the nation and human beings. Vanaspati has direct relationship with the well being of the society. Due to the pollution of natural environment and the destruction of forests many diseases crop up to ruin the nation. Only vanaspati with medicinal qualities may enhance the nature and cure diseases of human beings. Charak also warned *"when air, water and other elements of nature are spoiled, seasons start working against their routines or cycles, vegetation begins to ruin. This is most dangerous for both nation and its inhabitants."*

Ancient Indian literature such as Padmapurana well addressed the environmental pollution, the most environmental crisis we are facing today, in the following verse :

प्राणिनां प्राण हिंसायां ये नरा निरतः सदा
परनिन्दारता ये च ते वै निरयगामिनः
कृपाराम तडागानां प्रपानां च विदूषकाः।
सरसां चैव मेत्तारो नरा निरयगामिनः

Praninam prana himsayam ye nara nirat sada
parnindarta ye ca te yie nirayagamin
kuuparam tadagaanaa prapaanam ca vidhushaka
sarsaam chaiv mettaro nara niryagaaminah.

This means a person who is engaged in killing creatures, polluting wells, gardens, tanks and ponds certainly goes to hell. It tells us about the horizons of our understanding of the ecological systems.

Indian history reflects that people in India considered the life of a green tree is more valuable than their own lives. There were incidents where Indians sacrificed their lives in order to save the trees from axes. In 1730, in a small village of Rajasthan named 'Khejrali' the ruler had ordered cutting the Khejri (*Prosopis cineraria*) trees for baking lime for the construction of a fort. The local vishnois, basically belong to farming community, collectively protested this decision as they sensed the deprival of innumerable benefits of this multipurpose species if the trees were removed from their habitat. In their effort of protecting the life of green trees, 363 men, women and children lost their heads on the tree as the brutal soldiers axed their heads along the trees. The ruler of the state was then shocked by this sacrificial massacre and issued an order not to cut any more green trees in that region.[9] More than a couple of centuries later, once again in early 1970s the similar movement in the name of 'Chipko Andolan' saved the forest trees from large scale commercial tree felling programme in Garwal Himalayan region. Joint forest management is a successful environmental episode of recent time which reveals that Indians still hold of environmental safety traditional habits. Like wise the populace of India bear the valuable morals, ethics, customs and traditions which are proved to be in harmony with nature and its principles. Can't we adapt the merit of these ecological ethics for restoring our damaged ecosystems ?

References

1. R. K. Sinha and S. Upadhyay in *Environmental Crisis.* eds. R. K. Sinha and Sundarlal Bahuguna. INA Shri Publishers (1996).
2. Usha Govila, in *Indian Environment*, ed Pramod Singh, Ashish Publishing House, New Delhi (1992).
3. Mckim Marriott, *Village India*, (1961).
4. M. S. Randhawa, *Beautiful Trees and Gardens,* I C A R, New Delhi (1961).
5. Indira Gandhi, *Safeguarding Environment*, Wiley Eastern Limited, New Delhi (1992).
6. Stephen David, *India Today,* Sept. 30, (1996).
7. Banwar, *Pancavati - Indian Approach to Environment.* (Translated from Hindi by Asha Vohra) Shri Vinayaka Publications Delhi (1992).
8. R. K. Trivedi, *The Greening of Gujarat*, Director of Information, Govt. of Gujarat, Gandhinagar (1989).
9. Pramod Singh, (ed) in *Indian Environment*, Ashish Publishing House, New Delhi (1992).